P9-CRL-024

Our
Earliest
Ancestors

Björn Kurtén

Our Earliest Ancestors

Translated from the Swedish by Erik J. Friis

Illustrations by Viking Nyström

Columbia University Press
New York

Columbia University Press
New York Chichester, West Sussex

Library of Congress Cataloging-in-Publication Data
Kurtén, Björn.
[Våra äldsta förfäder. English]
Our earliest ancestors / Björn Kurtén ; translated from the Swedish by Erik J. Friis ; illustrations by Viking Nyström.
p. cm.
Includes bibliographical references and index.
ISBN 0–231–08061–1
1. Fossil man. 2. Human evolution. I. Title.
GN282.K8713 1993 93–6516
573.2—dc20 CIP

Casebound editions of
Columbia University Press
books are printed
on permanent and durable
acid-free paper.

Printed in the United States of America
c 10 9 8 7 6 5 4 3 2

Contents

Foreword

Headlines of the type "New Discovery Upsets All the Theories" are frequently seen. They create a vision of sparkling theoretical structures that suddenly collapse and crumble, eventually to be rebuilt from the ground up. But actually things are not that bad.

The broad outlines of human development have long been known. Nevertheless, one can say that the last two decades have seen quite a fantastic increase in our knowledge. Whereas in the past we had a mere handful of finds and discoveries about ancient humans, their number can now be reckoned in the thousands. Also, in other areas—molecular biology, geochronology, ecology, and in many other fields—science has provided new and in many cases

surprising answers to our questions. Thus research on our origins has been dynamic to a very high degree, which implies, of course, that practically every book on this topic will be somewhat outdated as soon as it is published. This book will not be able to avoid the same fate; in spite of that, it should be interesting to try to present an overview of what we know of this topic right now.

Trying to express oneself briefly means that one must pick and choose among the many facts and theories. I have tried to do this in an impartial way, but a great deal of material has inevitably been left out. Many of the controversies that are allotted space in the media are merely froth on the surface—for example, the current controversy between Leaky, Johanson, and others about which type of *Australopithecus* gave rise to our own genus Homo. It is a problem that cannot be resolved with the knowledge we possess today. I have preferred to concentrate here on matters that may be controversial but are important in their own right.

Each profession uses its own special terminology, and the reader will necessarily have to master certain basic concepts. Definitions of words and terms appear at the end of the book. I have attempted not to use unnecessary terms and expressions, but in one respect not much else can be done. Most extinct animals have no common names, and a direct translation of the Latin names becomes meaningless and even irritating (near human, partly human, southern ape, etc.), not least because in many cases the actual name is quite misleading. We will have to accept that the scientific name is only a label. As for myself, I can remember that when as a child I read books on paleontology I was fascinated by the magic inherent in names: *Brontosaurus* sounded as exotic and fired the imagination as much as Harun-al-Rashid did. Why not use *Ignacius, Proconsul,* and *Sivapithecus* just as well as Glumdalclitch, Charybdis, Pomuchelskopp, and Ura-Kaipa?

For those who wish to get more deeply into these problems, I have listed some of the literature on this subject at the end of the book. The list in turn may guide the reader to the primary and basic works.

Helsinki, February 1986

Our
Earliest
Ancestors

CHAPTER ONE

Humanity and Us

Myths about the origins of human beings are found in all cultures. In them humans appear quite ready-made: Ask and Embla, Adam and Eve. In a native American myth the first woman was borne by a mussel, and she became the ancestor of all humanity. Many of these stories have an inherent beauty. But reality is even more wonderful. Humanity is a result of a long development. "The long, magnificent story about the rise of man from among the dust of the stars is incomparably stranger, more awe-inspiring than any fable." These are the words of George G. Simpson, one of those scholars who have looked most deeply into the history of life on earth.

That knowledge has not been ours for very long. Not

until 1859, when Charles Darwin issued his book *The Origin of Species,* was the doctrine of evolution taken seriously by science. But we have known for a long time that animals physically very similar to humans exist. And it caused no great sensation that Linnaeus in his classification of the animal world included men and apes in the same order, "master animals" or Primates. It was probably because the classification was not considered an expression of kinship; all species of animals were then thought to have been created independently of each other and to be unchangeable.

The Order Primates belongs to the Class Mammalia—mammals. Other orders of mammals are, for instance, the whales, Cetacea; the rodents, Rodentia; and the beasts of prey, Carnivora. The mammals in turn are one of the many animal groups known as vertebrates, that is, animals with a firm, bony spine. Other groups include the fishes with bony skeletons, Osteichthyes; the reptiles, Reptilia; and the birds, Aves. This classification is thus a hierarchical system of taxonomic grouping (the classification Primates, for example, is a taxonomic one) in which each category becomes part of a higher category. Not until the doctrine of evolution was it clear that this hierarchy corresponds to different grades of real genetic relationships.

Traditionally, such classification is based on comparative anatomy. Every organ in our bodies has a counterpart among our relations in the animal kingdom. Thus the bones in our bodies are said to be homologous with bones in other vertebrates—as, for example, the lion. Developmentally speaking, this implies that if we follow the genealogical tables of ourselves and the lion far enough back in time (in this case about seventy million years), we will arrive at a common progenitor, and, second, that each homologous bone has appeared in its rightful place in every generation of the two lineal descendants. Incomplete homologies are due to evolution. The lion does not have a bone corresponding to our collarbone; it has retrogressed and become lost during the evolution of the felines.

The homologies are indeed numerous and detailed when we compare ourselves with the primates and above all with the great or anthropoid apes—the chimpanzees, gorillas, and orangutans. I will note some of these resemblances here. Primates tend to have large brains, and this is especially true of human beings and the great apes. Primates also tend to be "sight animals" rather than "smell animals." Our concept of the surrounding world is primarily based on our vision, while the world of a dog is based on smell. Among the higher primates, including all apes and human beings, the eyes are forward-directed so that their separate fields of vision coincide, and their assessment of distance (stereoscopic vision) becomes effective for the entire field of vision. Just like us, the great apes (like all the apes and monkeys of the Old World) have thirty-two teeth: in each half of the jaw they have

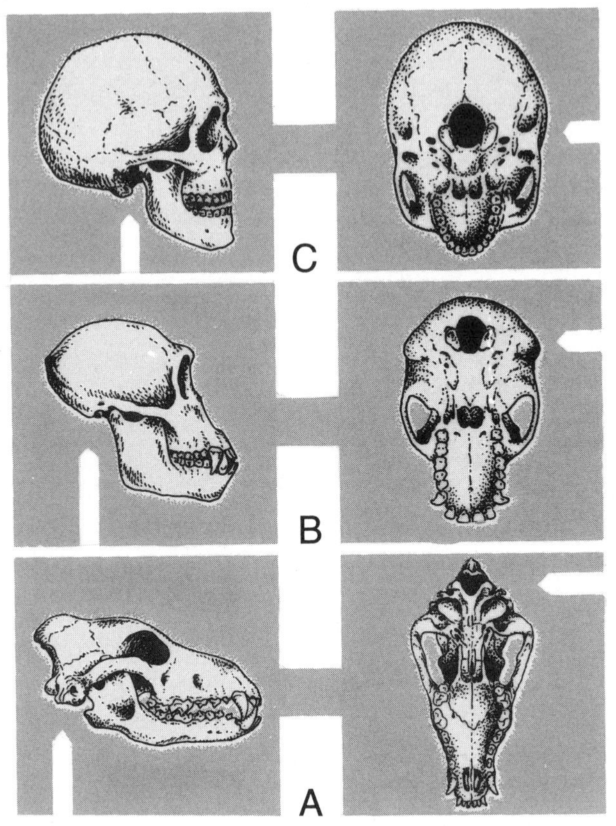

Figure 1.1 Skulls of (*A*) a beast of prey (wolf), (*B*) a primate (chimpanzee), and (*C*) a human being, which show the increasing dominance of the braincase and reduction in the size of the face and adjoining areas. Notice the location of the foramen magnum, which is flanked by the joint knobs that connect to the topmost vertebra. It is located directly in the center of the skull in human beings, who walk and stand upright, but in the wolf it is located at the back of the skull. In the chimpanzee, the foramen magnum is located slightly forward of that of the wolf. The chimpanzee has as many teeth as man, but the canines are larger. The wolf has a greater number of teeth (three incisors, one canine, four premolars, and two molars in each half of the upper jaw).

two front teeth, one canine tooth, two premolars, and three molars. The total number of teeth has been reduced from the time of the very earliest primates, which had forty-four teeth: three front teeth, one canine tooth, four premolars, and three molars in each half of the jaw.

On the other hand, our physique differs in important respects from that of the great apes. In accord with our upright stature, our pelvis, legs, and feet are constructed differently from those of the apes, and our vertebral column makes us swaybacked, which is not the case among the apes. We have shorter arms and fingers than the anthropoid apes, and our thumb is longer. Our heads are dominated by the braincase, while our faces are small and vertical. Apes have smaller braincases, and in their faces the jaws are dominant. Among the apes the canine teeth are large and cone-shaped; among human beings they are small and chisel-shaped. Humans have a thick layer of enamel around each molar, while the apes (except orangutans) have merely thin layers of enamel around each molar. I will mention other comparative features later. The differences between man and the great apes have led the systematists to assign them to different families within the primates (respectively, Hominidae and Pongidae).

The species that we know today may thus be regarded as the outermost twigs on a tree of evolution. Similar species are assigned to the same family, as, for instance, the Scandinavian lynx and the Canadian lynx, both of which are members of the lynx genus whose zoological name is just *Lynx*. When we compare them with the house cat, we see considerably less similarity, so the domestic cat is assigned to another genus, *Felis*. With this information we can conclude that the branching between *Lynx* and *Felis* probably occurred earlier than the branching of Scandinavian *Lynx lynx* from Canadian *Lynx canadensis*. Still, all three are actually so similar that they have been classified as members of the same family (the cat family, Felidae) within the order of the beasts of prey (Carnivora). Thus we now have two branches and a number of twigs on our tree of evolution. In a similar manner we can build the entire tree. As a result we obtain a "gestalt" showing the probable course of evolution through time; on the other hand, we do not obtain any "absolute" time scale. In order to obtain this we have to turn to paleontology.

During recent years research on the very molecules of life has led to a new method of determining the gestalt of evolution. Molecular biology compares the genes of different types and charts their dissimilarities. These have come into existence through mutations, which cause changes in the structures of DNA, deoxyribonucleic acid, the material of which genes consist. The differences in the

DNAs of closely related species are very small, but the differences grow larger the further removed the species are. The differences are due to mutations, above all the so-called point mutations, with the result that one of the organic molecules (bases) that connect the two DNA strings is exchanged for another. Such mutations accumulate in the DNA in the course of time; the longer the time, the more mutations. They thus function as a kind of "molecular clock."

As to actual details, the clock is not very exact. There is a big difference in the clock's "tempo" in various sectors of the DNA. In those sectors that directly affect the functioning of a protein, the clock goes slowly. In those sectors that do not affect such a function, the clock runs more quickly—it is evident that some mutations accumulated here do not have any practical effect. The theory about the molecular clock is based primarily on the study of such "neutral" mutations. (Whether a portion of DNA functions or not is determined by the so-called regulating genes, whose function may also be affected by mutations. Some of the differences between the various types may be dependent on regulating mutations.)

By analyzing the sequence of a given sector of DNA (or of a protein programmed for by DNA), one will consequently obtain a forklike representation, a gestalt of the development of such a tree, that may be compared to the one based on the classical method: comparative anatomy. It shows that the agreement between the two is very close. The molecular similarity is greater between humans and other members of the order of primates than between humans and members of any other group, especially when we compare them with the anthropoid apes. Similarly, various mammals are more like each other than they are like other animals, such as birds or reptiles.

As for details, however, there may be divergences. That may be because the molecular clock is less exact than we have wanted to believe. But it may also be because comparative anatomy has not provided the correct answer concerning the actual degree of relationship. We have a good example of that in our own case—our relationship vis-à-vis the great apes.

In the classical system humans are the solitary member of the family Hominidae (human beings). The great apes—gorillas, chimpanzees, and orangutans—belong to another family, Pongidae, and the small humanlike apes—the gibbons—belong to a third family, Hylobatidae. Together these three families make up a higher taxonomic group, a superfamily known as the Hominoidea. (Human beings are thus the only extant hominids; at the same time we, together with the gorillas, chimpanzees, orangutans, and the gibbons, are the contemporary hominoids.) All this can be expressed hierarchically, as in figure 1.2.

Molecular biology gives a somewhat different and probably more exact picture. According to its precepts, the earliest branching (A) occurred when the gibbons separated from the other groups, which thus include the great apes and human beings. In the next branching (B) the orangutan sets out on its own, while human, gorilla, and chimpanzee make a group of their own. Finally, that group separates into three (C). Here the situation is not quite clear, and various studies show somewhat different results. In some cases it seems as if humans initially sepa-

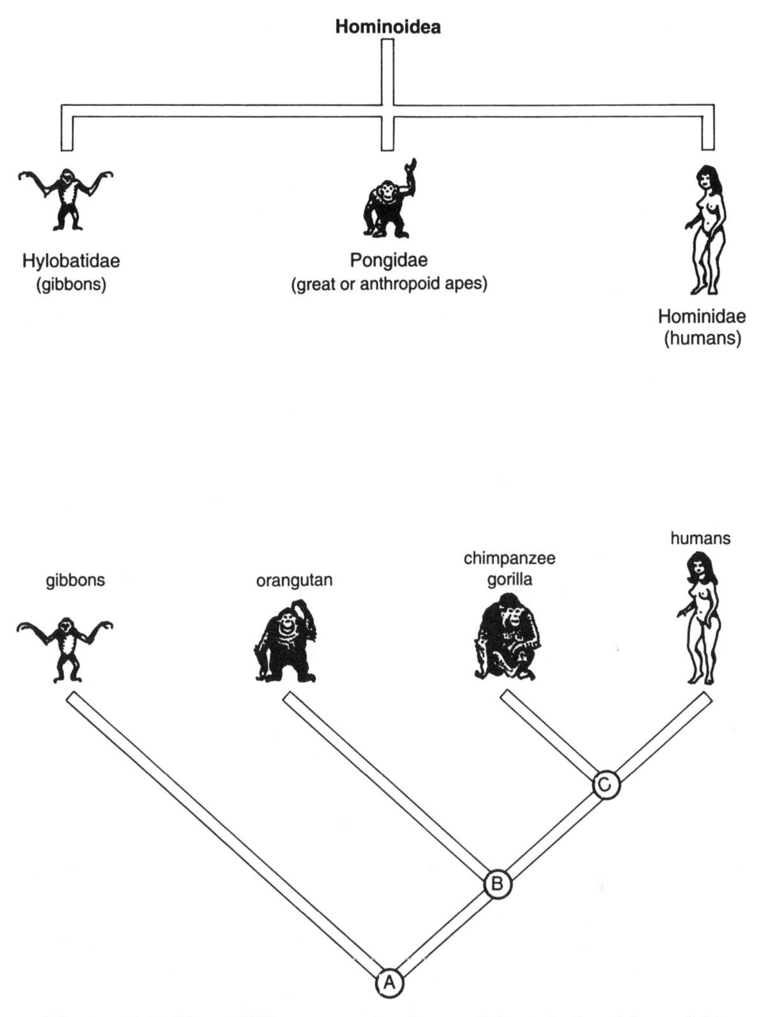

Figure 1.2 Two different conceptions of the relationships within the superfamily Hominoidea (i.e., humans and anthropoid apes). In the classical system (*top*) it is divided into the families Hylobatidae (gibbons), Pongidae (great or anthropoid apes), and Hominidae (humans). *Below:* a diagram showing the relationships according to molecular-biological investigations. See discussion in the text.

rated from the apes, and the gorillas are thought to have branched from the chimpanzees quite soon afterwards. Other studies indicate that the gorilla separated from the apes first, and humans and chimpanzees followed somewhat later.

In this connection we also have a time scale of sorts. The difference between humans and orangutans is almost twice as great as that between humans and chimpanzees. If the theory of the molecular clock holds true, it follows that the separation of the orangutan from the others (B) took place twice as long ago as the separation of the branch of chimpanzees (C). However, to obtain real time differences, we must calibrate the scale: we must determine the actual time difference for at least one of the points of separation, and that only paleontology can provide. The age of the other branching points may then be calculated.

We can thus realize that all members of the Hominoidea superfamily—those who are systematically classified as belonging to the families Hominidae, Pongidae, and Hylobatidae—have a common ancestral progenitor (the one represented by the point of separation A). For such a taxonomic group, paleontologists use the term *monophy-*

letic (that is, belonging to a single stock). The family Hylobatidae (the gibbons) is also monophyletic. But for the families Pongidae and Hominidae, the situation is entirely different. The taxonomic groups that descend from the same branching point are called sister groups, and in this case the sister groups consist of Hylobatidae on one side and Hominidae and Pongidae on the other. On the succeeding separation (B) two sister groups are created: one consists of the orangutan, the other of the human, the chimpanzee, and the gorilla.

The separation at B thus cuts through the family of the Pongidae, which indicates that this family is not monophyletic: one of the sister groups includes the Hominidae family and part of the Pongidae family; the other sister group consists of another part of the Pongidae. Pongidae is what is called a paraphyletic group.

We can avoid the problem by moving the chimpanzee and the gorilla from the Family Pongidae to the Family Hominidae. We then have two monophyletic families: the Hominidae, which includes the human, the gorilla, and the chimpanzee, and Pongidae, in which the orangutan is all alone.

This is a so-called phylogenetic classification. Its aim is to represent as exactly as possible the evolutionary process and its genetic connections. It does not allow any paraphyletic taxonomic groups.

But it is not devoid of problems. As we see, it places the chimpanzee and the gorilla with the human in the same family, Hominidae. This reflects the fact that their DNA is much like that of humans. Still, they cannot possibly be considered human beings. Despite all similarities they differ from humans both anatomically and also in behavior. (The dissimilarities are thought to be due, among other things, to the regulator genes.) Pure phylogenetic systematics has one weakness: it pays no regard to the organism's "life niche," to evolutionary breakthroughs and the acquiring of new lifestyles. One just can't ignore the fact that the chimpanzee, the gorilla, and the orangutan are apes and are closely related anatomically and in behavior, while the human has taken a completely different evolutionary path.

Thus we have a choice between two different types of classification. The phylogenetic is concerned merely with the gestalt of evolution and thus does not allow for any paraphyletic taxonomic groups. The other type may be characterized as "eclectic" and attempts naturally enough to take account of phylogenics, but it also considers the type of organization and level and permits certain types of paraphyletic entities. Both have certain advantages and weaknesses, and whichever is preferred depends on one's personal philosophy.

No matter which classification system we use, "the last common ancestor" is an essential concept. In the diagram, A represents the last common ancestor of Hominoids; B represents the last common ancestor of orangutans and human beings; and C represents the last common ancestor of gorillas, chimpanzees, and humans. Molecular biology tells us about the existence and relative ages of these ancestral groups but cannot tell us about the way they looked and how they lived. Nor can we "reconstruct" any of these ancestors by applying a proportional mean between man and the chimpanzee (being an approximation

to ancestor C). Such a proportional mean would, for example, have a brain size equidistant between that of humans and that of the chimpanzee; this would imply that the size of the chimpanzee's brain has decreased since stage C. However, it is more probable that the brains of the chimpanzee and of the human have grown in size ever since stage C—that of man at a faster pace.

The only way we can learn more about all this is to use the testimony of fossils; only they can help us obtain a picture of our earliest ancestors. At the same time they will provide a firm time scale in which the tree of evolution and its many branches can take root.

CHAPTER TWO

A Long Story, Briefly Told

The universe came into being between ten and twenty billion years ago; according to one theory it is close to fifteen billion years old. The solar system and the earth are considerably younger but still of an impressive age: 4.5 to 5 billion years. It is clear that life already existed on earth a billion (1,000 million) years after the origin of the earth.

The history of evolution goes back to the first sign of life on earth—thus for the sake of completeness we ought to look at developments long preceding the appearance of the primates. In terms of the time involved, this means 98 percent of our entire prehistory, while primates have existed in a mere 2 percent of that time and man a mere 0.004 percent.

The first signs of organic life have been found in sediments deposited on the sea bottom 3,500 million years ago. They are the remains of bacteria and cyanophytes ("blue-green algae"), the simplest, least complex organisms we know of. Developments during the next two billion years were exceedingly slow. One reason may have been that organisms at this level multiply by asexual reproduction; consequently, each individual has only one parent, and there is no combination of gene mutations such as may have occurred from the union of two dissimilar individuals. Suppose that ten gene mutations take place in such an asexual population: the result is eleven dissimilar gene types (the original one plus the ten new ones). On the other hand, ten mutations in a population with sexual reproduction, where all the variants may be combined and recombined, may result in 3^{10} or almost sixty thousand different gene types. This unheard-of increase in the varied material available in natural selection makes possible a radical increase in the pace of evolution.

This occurred when the earliest organisms with a nucleus appeared approximately 1.5 billion years ago. They were still one-celled organisms but were the first examples of so-called conjugation, that is, sexual merger preceding the process of division. Among these organisms are primary producers (those that contain chlorophyll and build up organic matter through photosynthesis—thus being mainly plants) and consumers (animals).

About 600 million years ago, the first multicelled animals made their appearance. Thus began the so-called Phanerozoic eon—a long series of geologic periods the history of which is documented by a rich fossil record.

Figure 2.1 shows how the eon has been divided. It should be noted that in figure 2.1 time runs upward—in other words, the more deeply we dig, the further back in time we go, exactly as with geological deposits. (All diagrams with time indicated in the opposite direction will seem baroque and amateurish to a geologist.)

The periods (which have an average length of approximately fifty million years) contain three eras: the Paleozoic (the "ancient time" of life), the Mesozoic (the "middle period" of life), and the Cenozoic (the "new period" of life).

The Paleozoic era started with the Cambrian period, the first to leave behind rich fossil material with multicelled animals. (Multicelled animals are also represented among late pre-Cambrian animals, but they are not completely known, and many of them seem to represent completely extinct lines of development.) Practically all the main groups or types of animals whose physical makeup includes a hard, bony skeleton and that have a chance of being preserved as a fossil appeared during the sixty to seventy million years that the Cambrian lasted. These animals include the sponges, cnidarians, annelids, molluscs (probably as early as in the Cambrian period), arthropods, echinoderms, as well as some now completely extinct kinds.

By the middle Cambrian we find a fossil of an animal

Figure 2.1 The history of the earth from its creation approxiomately 4.5 billion years ago. *Facing:* the Phanerozoic eon, which corresponds to the diagram on the right. The history of the earth in the Phanerozoic eon is divided into three eras—Paleozoic, Mesozoic and Cenozoic—and they in turn into periods, from Cambrian to Quaternary.

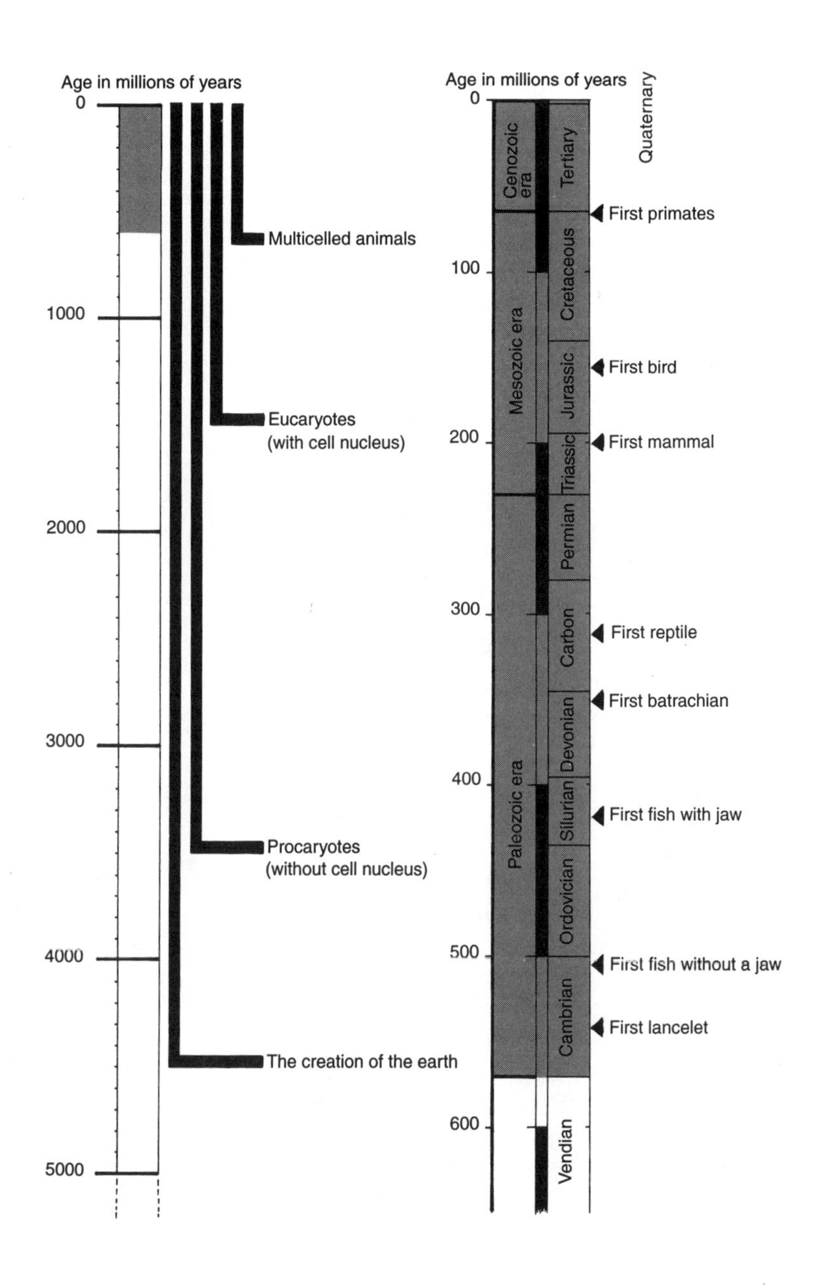

Age in millions of years
0
1000
2000
3000
4000
5000
Multicelled animals
Eucaryotes
(with cell nucleus)
Procaryotes
(without cell nucleus)
The creation of the earth
Age in millions of years
0
100
200
300
400
500
600
Quaternary
Cenozoic era
Mesozoic era
Paleozoic era
Tertiary
Cretaceous
Jurassic
Triassic
Permian
Carbon
Devonian
Silurian
Ordovician
Cambrian
Vendian
First primates
First bird
First mammal
First reptile
First batrachian
First fish with jaw
First fish without a jaw
First lancelet

centrally important in our own early history. The animal, named *Pikaia,* was part of a group of fossil animals found in British Columbia, Canada, where local conditions (stagnant water, in which decay was incomplete because of a lack of oxygen) preserved the organisms in fine-grained sediments. *Pikaia* was a lancelet, closely related to the present-day type. The lancelet differs from actual vertebrate animals in its lack of both an interior skeleton and a real head, but otherwise the similarities are so great (for example, the presence of a dorsal nerve cord, notochord, gill openings, an intestinal canal located underneath the nerve cord, segmented muscles) that it is considered the nearest relative of the vertebrates. The vertebrates and the lancelet probably had an original common ancestor in the Cambrian period. This ancestor may have been *Pikaia* or a closely related type.

The earliest genuine vertebrates appear in strata that date from the upper Cambrian period. Thus they are approximately 500 million years old (while *Pikaia* lived thirty to forty million years earlier). They resemble the lancelets in that the mouth lacks jaws and that the slits connected with the gills are well developed and function

---

Figure 2.2 From lancelet to mammals. The pictured stages are *Pikaia* (lancelet, Cambrian), *Pteraspis* (agnatha, Devonian), *Osteolepis* (lobed fin, Devonian), *Ichthyostega* (amphibian, Devonian), *Hylonomus* (reptile, Carbon), *Varanosaurus* (early mammal-reptile, Permian), *Thrinaxodon* (advanced mammal-reptile, Triassic), jaw of *Spalacotherium* (primitive mammal, Jurassic), and *Zalambdalestes* (insect eater, Cretaceous). *Right,* some present-day descendants of the fossil varieties: lancelet, slime eel, lobed-fin *Latimeria,* frog, lizard.

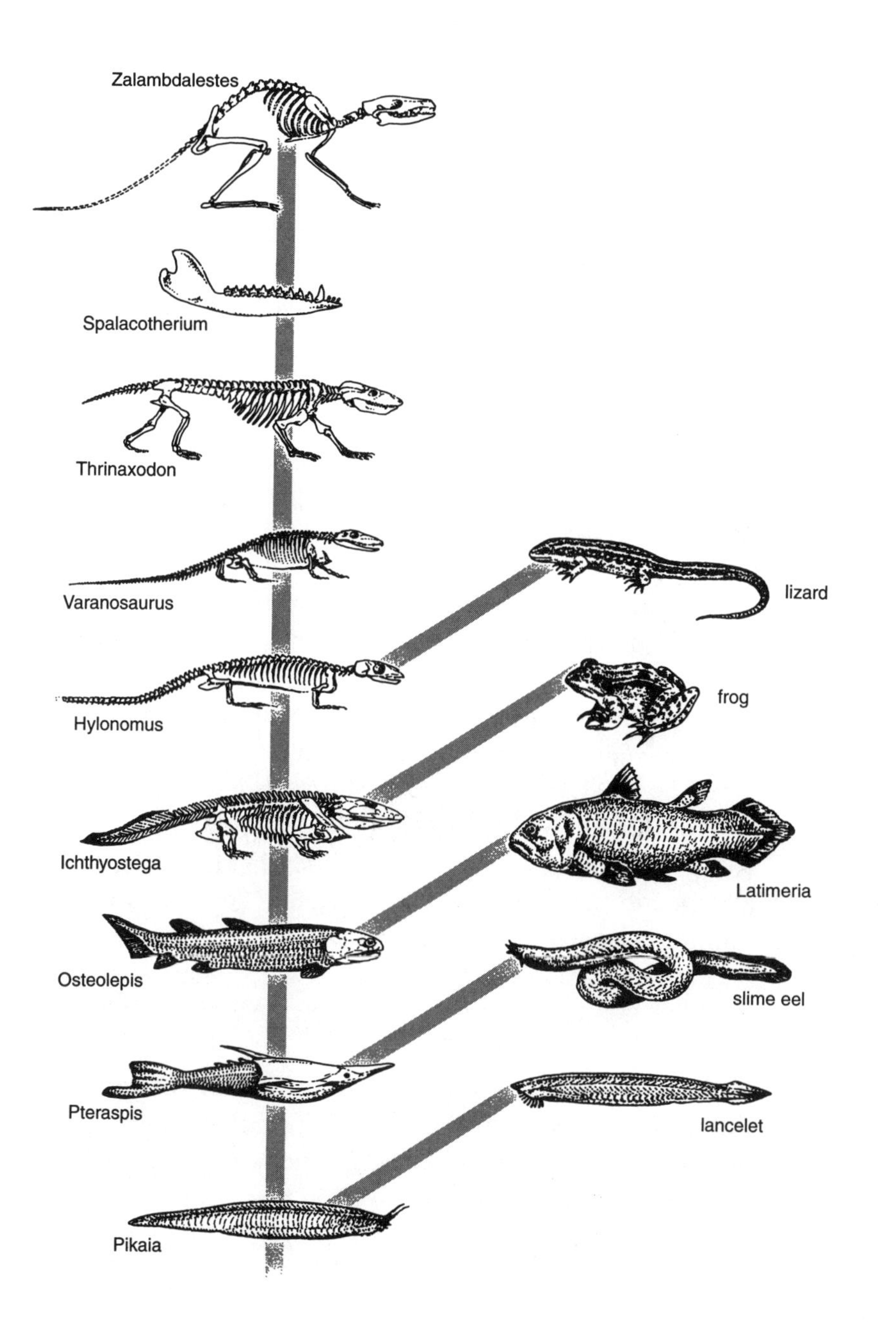
Zalambdalestes
Spalacotherium
Thrinaxodon
Varanosaurus
lizard
Hylonomus
frog
Ichthyostega
Latimeria
Osteolepis
slime eel
Pteraspis
lancelet
Pikaia

as the organs for obtaining nourishment (by straining particles of food from the water that flows through), but in contrast to the lancelets they have a strong ossified skeleton and a well-developed head with organs for sight, smell, and hearing. These fish are called agnathans (jawless) and include many types that lived into the Devonian period—and they still exist in the form of cyclostomes, lampreys, and hagfish.

The next stage in this development was the appearance of real jaws, a change that took place as early as the middle Silurian period and led to an unprecedented flourishing of marine fauna during the Devonian period, which with good reason has been called "the age of fishes." In oceans, lakes, and rivers there lived agnathans, cartilaginous fishes (including sharks), and bony fishes of many types in addition to the numerous but now extinct so-called placoderms (armored fishes).

The types of fish that probably hold the greatest interest from our viewpoint are the freshwater fish that have lobed fins and are known as crossopterygians. Like contemporary lungfishes (which exist today basically unchanged) they had a swimming bladder that functioned as a lung, and they could thus breathe air. They had powerful, muscular pairs of breast and belly fins with an inner skeleton consisting of several bones, which made the fins flexible and useful for support during overland travel. The fact that the fishes with lobed fins actually crawled on land is proved by fossil footprints from the middle Devonian period. It is thus from this time, 370 million years ago, that we invaded the continents. By then other organisms had already established themselves on firm land. Forerunners of ferns, horsetails, and club mosses as well as coniferous trees formed thin forests in moist soil, with wingless insects and scorpions living in them. The first vertebrates living solely on shore, which Professor Erik Jarvik in Stockholm has called the "four-legged fishes," arrived toward the end of the Devonian period. These animals can be considered primitive amphibians, in which the heritage from fish is extant in the form of a rudimentary gill cover and part of the tail fin. With that we have definitively climbed up on land—a fact confirmed by the fossil prints dating from the late Devonian period.

Amphibians are four-footed animals that breathe air but are still dependent on water for reproduction: the roe are laid in water and then fertilized by the male (exterior fertilization). During the Carboniferous period, which has been called the "era of the amphibians," the batrachians (Amphibia) were indeed the dominant land animals. But by the middle Carboniferous period, about 310 million years ago, evolution had advanced as far as the reptiles, which have internal fertilization and lay their shell-like eggs with nourishing yolks on dry land. Thus we have finally freed ourselves from existence in the water.

From the original Carboniferous forms, the reptiles developed into two large groups, which during the Permian period became the lords of the land: the era of the amphibians was over, and that of the reptiles was about to begin. It was to last a very long time: 200 million years, to the end of the Cretaceous period.

Of the two types of reptiles, one branched off to become the different reptiles of the present day: tortoises and turtles, crocodiles, snakes, and lizards. It gave rise to

all the dominant reptiles of the Jurassic and Cretaceous periods: dinosaurs, flying lizards, and tremendous sea lizards of various types. The birds—besides the crocodiles—are the sole descendants of these so-called dominant reptiles.

The other group is made up of the mammal-like reptiles (Synapsida) that appeared as early as the late Carboniferous period and, of the Permian and Triassic periods, played an important role as in the faunas. However, the Synapsida lost the competition with the dominant reptiles and in the middle of the Triassic period gave up their dominant position. But they didn't become extinct. During the transition to the late Triassic period—approximately 205 million years ago—they had given rise to the first mammals.

The origin of mammals is likely one of the most remarkable examples of evolution documented in fossils, especially since fossils of neither synapsid reptiles nor early mammals are often found in layers from the late Triassic period; they were unnoticed during their lifetime and remained so after death. But we do have evolutionary series in which it is exceedingly difficult to draw a line and say: from this year on they were mammals. Such a dividing line used to identify mammals is the number of bones in the lower jaw. Reptiles have several bones there, the mammals only one. But even that threshold has evidently been crossed within more than one lineage.

From a skeletal and anatomic viewpoint, these animals from the late Triassic period are classed as mammals—naturally enough, we cannot determine whether all the characteristics of the mammals had already appeared. Mammals are viviparous, that is, they bear live young, but the monotremes one finds in Australia lay eggs, and possibly all the earliest mammals did the same. Young mammals obtain nourishment by suckling their mothers; mammals have hearts with four chambers and double circulation of the blood; they have warm blood, but there are exceptions; they have hairy coats or furs (less in certain types). In all these respects mammals differ from the typical reptiles. However, some of these characteristics were beginning to develop even among the Synapsida. Small hollows in the upper jaws may have been attachment points for whiskers, which indicates that the skin was covered by hair. Hair may indicate warm-bloodedness, and the same is indicated by convoluted nostrils, in which the air breathed in is warmed. Other signs of increased metabolism are the division of the bite into incisors, canines, and molars with different functions, and the appearance of a secondary palate roof below which inhaled air is led directly to the trachea without passing the oral cavity.

Among early amphibians and reptiles the front and hind legs stick straight out from the body. The Synapsida, on the other hand, have legs that tend to be turned in under the body, raising it up from the ground. Thus, instead of having to crawl, the advanced Synapsida were able to walk and run like mammals.

For a long time, the development of mammals during the Jurassic and Cretaceous periods was not well known—and it covers two-thirds of their history. New collection methods have provided a comprehensive fossil record, and ongoing research is intensive. Nevertheless, the gaps are still far from small.

By the Triassic period the mammals seem to have divided into two branches. One, Prototheria, includes various extinct groups; the monotremes of mammals that are native to Australia are regarded as the present-day representatives of the Prototheria. A recent find of an early mammal (from the lower Cretaceous period, from Australia) is thought to belong to the other branch, Theria, which in time gained the upper hand. This line of development leads to present-day marsupials and the "higher" or "placental mammals." Early marsupials and placental mammals appeared as early as the middle Cretaceous period; toward the end of the Cretaceous period Theria were the dominant mammals, while most of the members of Prototheria had become extinct.

During the Mesozoic Era the mammals lived in the background in a world dominated by dinosaurs and other large reptiles. These mammals were all rather small—the biggest may have weighed ten kilos, and most were much smaller. I have mentioned some of the characteristics that differentiated them from the contemporary lizards, but not the most important one, namely the brain. Already in the Mesozoic mammals the brain had grown larger than in reptiles of corresponding size, and this development continued. This growing improvement in the central nervous system was probably one of the most important characteristics that allowed mammals to survive and assume ascendancy when the fateful hour struck for the dinosaurs at the end of the Cretaceous period.

While certain Prototheria developed into highly specialized plant eaters, the Mesozoic Theria seem in general to have been omnivorous animals or small beasts of prey—a type of insect eater. Many Mesozoic mammals were probably tree climbers, and that probably even held true for the ancestors of the primates.

The Cretaceous period ended about sixty-seven million years ago, accompanied by a dramatic change of scene, which is still the subject of debate. All dinosaurs died out, as did many other types of animals that until then had been numerous and had displayed many sizes and shapes. The time span during which the extinction took place is still far from clear. Was it connected with a catastrophe taking a year or less, or was it a gradual process over a thousand years or more? A theory currently popular has it that the earth was hit by a giant meteorite, resulting in the atmosphere becoming so muddy and polluted because of the dust thrown into the air that an "Egyptian darkness" set in, lasting perhaps a year or more. This could have led to the death of plant life and consequently a major part of

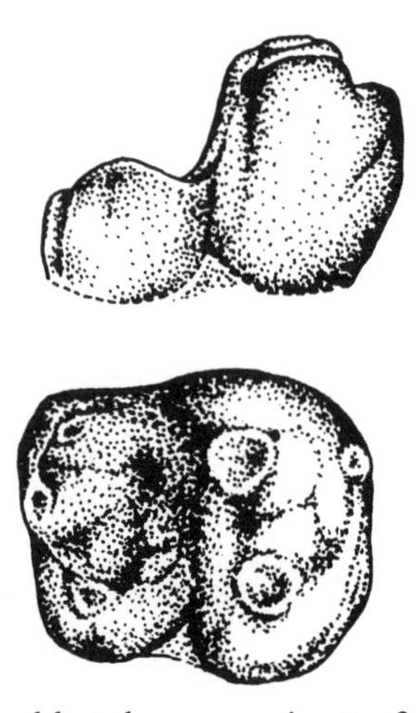

Figure 2.3 The oldest known primate fossil is this tooth of *Purgatorius* from the upper Cretaceous period in North America. We know that it is *Purgatorius* because a corresponding tooth belonging to the same family has been found in somewhat younger sediments from the early Paleocene, which provided a larger number of specimens. The crowns of the cusps are somewhat worn (from chewing), but we can still see that they are higher and more pointed than those of typical primates, an inheritance from ancestors that were insect eaters.

the animals. Nevertheless, no change in flora took place at the end of the Cretaceous period—such a change had already occurred about forty million years earlier. Besides, a great number of animal types—even reptiles—survived the supposed catastrophe.

Other theories, less spectacular but probably much closer to the truth, hold that changes in the climate caused by geologic processes gradually influenced the environment so that the large reptiles were affected (it has been shown, for instance, that dinosaur eggs became ever thinner toward the end of the Cretaceous period, a result perhaps of lowering temperatures).

The problem of the mass deaths at the end of the Cretaceous period is thus unsolved, but we can at least maintain that they did not affect the primates. For right before the end of the Cretaceous period the earliest of the primates came on stage. It was a small, primordial being called *Purgatorius* (named for the place where it was found, Purgatory Hill in Montana). And we again meet the same *Purgatorius* in the same area *after* the dinosaurs had died and the world had entered a new time period: the Tertiary period.

CHAPTER THREE

Prosimians: Types and Collateral Branches

Purgatorius, the earliest known primate, was about the size of a squirrel. Unfortunately, the only fossil remains we have are teeth and parts of its jaws. But even this unassuming material provides certain clues as to both its descent and way of life. Generally speaking, the shape of its teeth seems like something between that of typical early primates and insectivores (belonging to the group currently represented by such animals as the hedgehogs and the shrew mice). The molars still have high cusps (crown points), which among the insectivores cut through the hard shell of their prey. We have learned that *Purgatorius* also ate fruits and/or leaves. Thus it seems as if *Purgatorius*—like so many other primates—was an omnivorous animal rather than one that specialized in certain types of food. Among later primates the cusps become lower and blunter, the norm among animals with a variegated diet. *Purgatorius* thus represents a link between the primates and their forerunners among the insectivores.

From *Purgatorius* or from some closely related type, a multitude of different varieties developed during the Paleocene period, the first epoch of the Tertiary period (approximately sixty-seven to fifty-five million years ago). They all belong to the subgroup Plesiadapiformes, which is completely extinct. We may call them primitive prosimians. They were small—the same size as rats and squirrels—and probably were forest animals, often perhaps living in trees. Their fossils have been found in Europe, Asia, and North America, which during the early Tertiary period had a tropical or subtropical climate even in North European latitudes.

Otherwise they exhibited a great diversity. More than sixty kinds are known, distributed over twenty genera and five families. The great variety among these animals reached its zenith during the late Paleocene period. At the end of that epoch most of the Plesiadapiformes died out, but some types survived past the early Eocene period (fifty-five to fifty-two million years ago). And the type known as *Phenacolemur* existed as late as the middle Eocene period (forty-eight million years before the present day), while the type known as *Ignacius* actually lived into the late Eocene period (thirty-nine million years ago). It is noteworthy how many (as, for example, the two types just mentioned) developed a set of rodentlike teeth with powerful gnawing front teeth and grinding molars. Some of them developed enormous saw-toothed lower premolars (similar ones are found among certain kangaroos and many other living and extinct animals), which quite possibly were used to chew fibrous food—as, for example, cambium tissue underneath the bark of trees.

Most of the Plesiadapiformes are so similar to rodents that one hypothesis about rodents postulates that both the orders Rodentia (rats, field mice, beavers, etc.) and Lagomorpha (harelike animals) descended from Plesiadapiformes. This theory has received cautious support through certain investigations conducted by molecular biologists. If it holds true, we have indeed received unexpected company.

It is clear that none of these specialized, rodentlike prosimians can have been the original progenitors of the higher primates, whose origins must be sought among the primordial types of Plesiadapiformes living in the early Paleocene period (among them *Purgatorius* itself).

The first members of a new suborder, Strepsirhini, appeared at the beginning of the Eocene period (fifty-five million years ago); this group assumed dominance during the following millions of years. All known Strepsirhini from the early Tertiary period belong to the family Adapidae, which mainly died out by the end of the Eocene period (thirty-seven million years ago). One type survived an additional million years in France, and an Adapid that lived as late as eleven million years ago has been found in India.

While most of the Plesiadapiformes were similar to rodents, this specialized type never appears among the Adapidae. The Adapidae had small incisors, usually strong

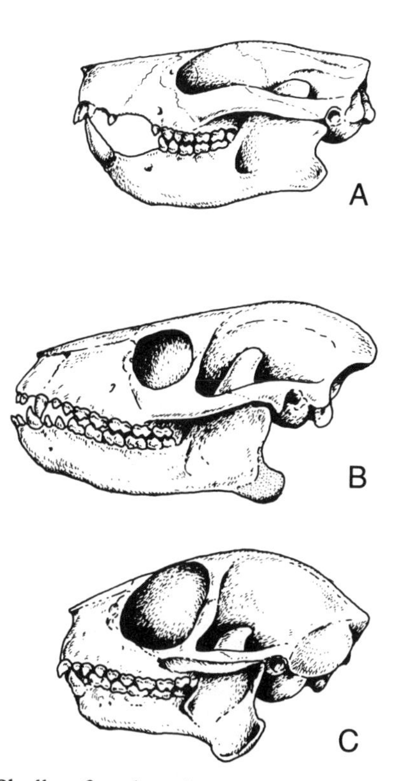

Figure 3.1 Skulls of early primates: (*A*) *Plesiadapis* (Plesiadapiformes); (*B*) *Notharctus* (Adapidae); (*C*) *Necrolemur* (Omomyidae). In (*A*), note the eye socket, which opens towards the rear, and the rodentlike tooth system; in (*B*) and (*C*), note the rows of teeth, which are close together, the ring of bone around the eye, and the increasing size of the brain.

and powerful canines, with no spaces between the teeth. The structure of the eyesockets represent an important advance. Among the Plesiadapiformes, the eye socket opens toward the back in the region of the temples, while behind each eye the Adapidae has a bony ridge enclosing

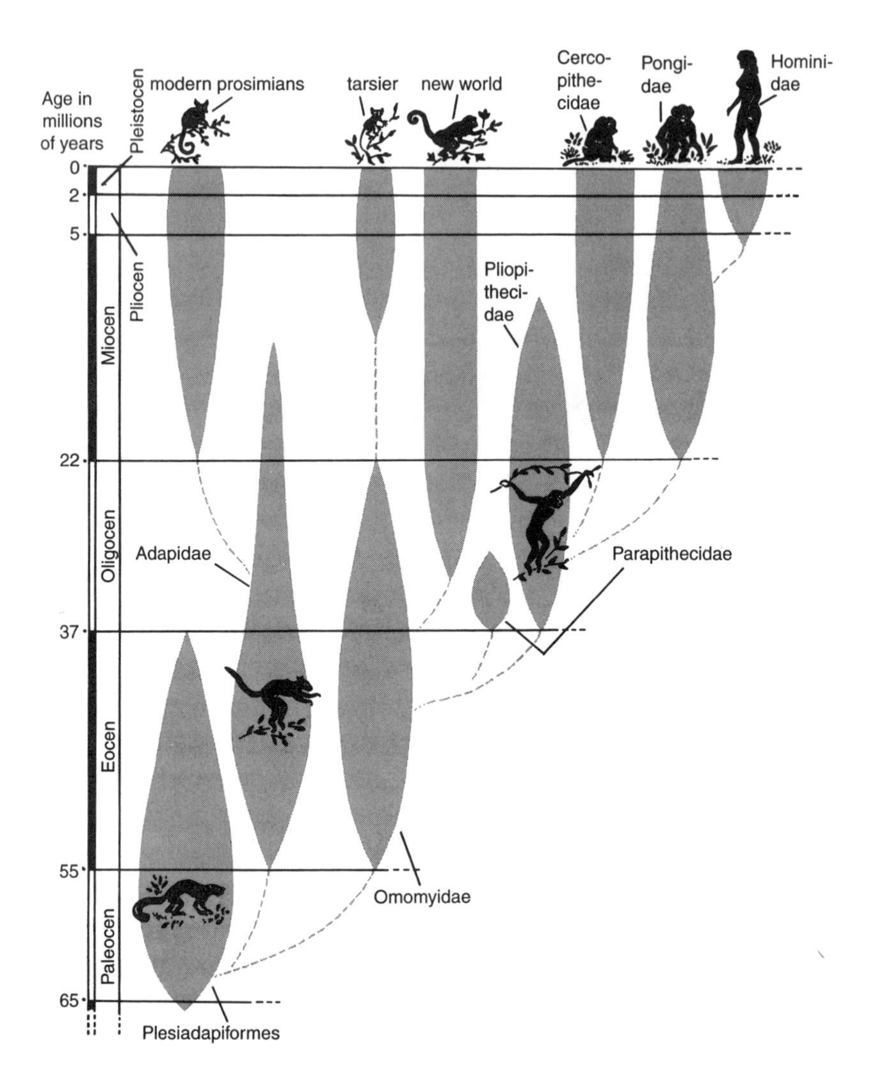

Age in millions of years
Pleistocen
Pliocen
Miocen
Oligocen
Eocen
Paleocen
0
2
5
22
37
55
65
modern prosimians
tarsier
new world
Cerco-pithe-cidae
Pongi-dae
Homini-dae
Pliopi-theci-dae
Adapidae
Parapithecidae
Omomyidae
Plesiadapiformes

the entire eye. This caused their eyes to be more forward-directed than those of the Paleocene types.

In general the Adapidae were larger than the Plesiadapiformes. Some of them became as big as our housecats or even a bit bigger. They were typical tree climbers, long-tailed animals probably closely resembling their present-day descendants, the lemurs. The main characteristic separating today's Strepsirhini prosimians from the Adapidae is the development of the incisors and canines into a "comb," which probably evolved for cleaning fur (and in certain cases still serves the same purpose).

The family Adapidae comprises approximately thirty different types, distributed among at least seventeen genera—the finds originated in Europe, North Africa, Asia, and North America. What we know about them surely is merely the tip of the iceberg: during the eighteen to nineteen million years of the Eocene period a great number of different varieties must have existed.

The Strepsirhini are currently represented by the lemurs, *Indridae, Galagos,* and the aye-aye (the finger animal). They all have made their home in the tropics of the Old World; today these prosimians are found above all on

---

Figure 3.2 Summary of the development of primates during the Cenozoic era. The periods (Tertiary and Quaternary) are divided into epochs from Paleocene to Pleistocene (Ice Age). The present period (Holocene) is too brief to be included. The symbols show the various lengths of time the different primate groups existed (for instance, Plesiadapiformes existed from approximately thirty-nine to sixty-seven million years before our time), while the width of the symbols indicates the relative abundance of types.

the isolated island of Madagascar. They are without a doubt direct descendants of the Adapidae.

Did the progenitors of the Adapidae also belong to the higher primates—apes and humans? Many researchers have been of that opinion. But today only a few take it seriously. By the beginning of the Eocene period—at the same time as the first Adapidae—members of another family appear on the stage, namely the Omomyidae, which fit much better into that role. This family probably included many more subtypes than did Adapidae (as many as forty-five subtypes are known, distributed among twenty-seven genera, from the Eocene period; a few genera also existed in the Oligocene period).

A present-day descendant of the Omomyidae is the *Tarsius* or the "ghost animal," which lives in the Philippines, Borneo, Sumatra, and Celebes. This miniature primate, no bigger than a rat, looks straight ahead with very large eyes. It is nocturnal, lives in trees, and preys on insects and small vertebrates. Its name alludes to its greatly elongated tarsus, or distal part of the foot, which enables it to make long jumps from one branch to another. *Tarsius* has long been considered an intermediary link between the prosimians and the actual apes. But it is the result of a long developmental history, with its Eocene antecedents, the Omomyidae, forming the real link.

Among the Omomyidae we find many of the advanced characteristics of *Tarsius,* such as a tendency for the eyes to look straight ahead, but not *Tarsius*'s unique characteristics, such as the lengthened tarsal section. Most of the Omomyidae had two incisors, one canine, three premolars, and three molars in each half of the jaw, exactly like present-day apes of South and Central America. Those too are probably descendants of the Omomyidae. In *Tarsius,* a front tooth in the lower jaw has disappeared. In the apes of the Old World as well as in humans, on the other hand, it is the premolars that have been reduced in number, from three to two.

During the early Tertiary period—the Paleocene and Eocene epochs, about thirty million years—the forests in Europe, Asia, and North America were teeming with prosimians. (Probably in Africa also, where they are hardly known, since up to now hardly any fossil-bearing strata of that age have been discovered there.) Present-day prosimians are denizens solely of tropical forests, and the same seems to have been true of earlier types. Flora originating in London in the Eocene period remind us most closely of the flora of the present-day Malacca peninsula. Other floras in Asia and North America tell the same story—about a time when the northern continents had tropical or subtropical climates. It has been estimated that the annual average temperature in Europe during the Eocene period was approximately eleven degrees (C) higher than it is today.

We know only a fraction of the many Plesiadipidae, Adapidae, and Omomyidae that existed in Tertiary. It is therefore highly improbable that the fossils we know belonged in the lineage that led to the higher primates and to ourselves. If a complete lineage were available, we would probably be able to classify the earliest members as Plesiadapidae and the later ones as Omomyidae. But from our point of view, most of those we have been looking at make evolutionary collateral branches. And we probably need to concede that the same is true of most of the fossil history of the primates. Finds of fossil primates are very rare. One reason is that their usual milieu—the forest—rarely creates the conditions in which fossils are easily preserved. Fossils accumulate in sedimentary channels or basins some distance from forests (for example, in rivers, lakes, seashores) or in caves. Another reason is that the primates discussed so far, namely the prosimians, are small and have brittle skeletons that easily shatter before they end up in a protective sediment. Also, primates are rarely represented by large numbers of individuals, and they generally are too smart to become a fossil—as, for instance, by falling into the water and drowning.

The almost explosive development of the prosimians in the early Tertiary period is an example of *adaptive radiation:* the development of a multitude of dissimilar types, adapted to various milieus and ways of life, from a single primitive type. On a smaller scale the same occurred within each of the three groups that we have considered: Plesiadapidae, Adapidae, and Omomyidae. And we encounter the same phenomenon time after time when we get to the following stage in our developmental history: the story of the great apes.

CHAPTER FOUR

Africa: The Original Home?

The Oligocene period lasted from about thirty-seven million to about twenty-five million years before our time. An enormous mass of land mammals from the Oligocene period has been found in Europe, Asia, and North America. Still, no fossil of a single ape has been found on these continents. All the Oligocene apes lived in either South America or Africa.

We do not have to concern ourselves here with the South American ones, since they are not as closely related to us. They belong to a group (a so-called infraorder) called Platyrrhini, or "broadnoses," while the apes of the Old World and we ourselves are Catarrhini, "narrow-

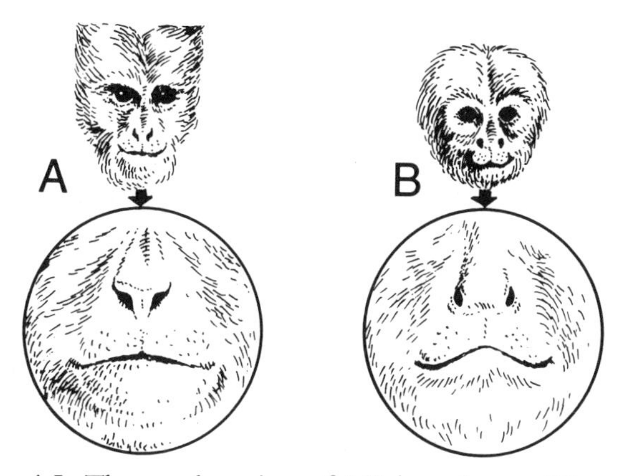

Figure 4.1 The nasal section of (*A*) broadnoses (Platyrrhini), and (*B*) narrow-noses (Catarrhini); the pictures show a howler monkey and a macaque. The broadnoses include the New World monkeys. The Old World monkeys, apes and human beings are narrow-noses.

noses." The names indicate an important anatomical difference, but there are many others.

About one hundred kilometers southwest of Cairo, in the province of Fayum, the high cliff Djebel el-Qatrani rises above the desert landscape, with a view toward the south and the Qarun Lake. The mountain is a witness to the process of erosion—somewhat like the mesas of the southwestern U.S.—but it has been protected from being leveled by a resistant basalt outer layer. Below the basalt is a series of layers with fossils, and we derive all the information we have about our narrow-nose Oligocene ancestors from here.

The bottom layers date from the Eocene period and are what may be called marine. Formed from the bottom of the Mediterranean, which constantly inundated the entire area, they contain the remains of whales, sea cows, crocodiles, and other marine animals. The overlying sediments contain fossils of land or freshwater animals that lived in the marshy Nile delta during the early Oligocene period. This fauna exhibits a great number of types, and an interesting fact is that most animal types represented here are *endemic* to Africa in the Oligocene period—they are not found anywhere else. These animals include, for example, early mastodons (a kind of elephant), relatives of the so-called rock badgers, as well as extinct varieties of apes. It is quite evident that Africa was isolated from Europe and Asia during the Oligocene period; none of the animals referred to here have existed in Eurasia. Practically all the typical Eurasian varieties are likewise missing in Africa—rhinoceroses, piglike animals, and real beasts of prey.

A lava flow, which seals the layers at the top, can be dated through the use of radiometry—a method used to investigate certain radioactive elements. By applying this method a "definite" geologic chronology has been constructed. An earlier attempt at dating indicated that the lava originated at the end of the Oligocene period, but more recent research shows that lava eruptions began as early as thirty-one million years ago. The Oligocene fossils from Fayum are consequently between thirty-five and thirty-one million years old.

Two main types of fossil narrow-noses have been found at Fayum. One is the Parapithecidae (consisting of two genera and four species), very small apes, one type of which is the most common fossil at Fayum. They may be

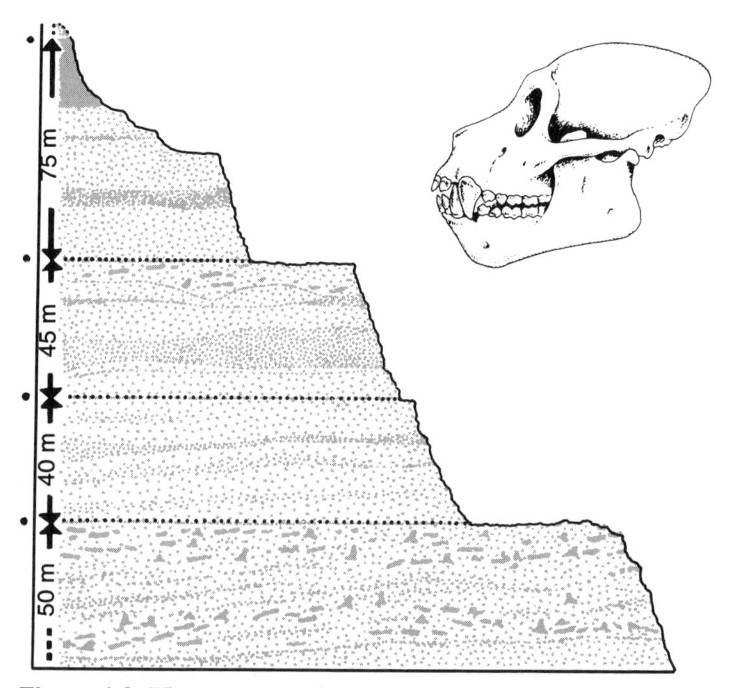

Figure 4.2 The sequence of layers at Djebel el-Qatrani in Fayum, Egypt. At the base of the hill are marine layers from the Eocene period. The hill itself was built up by Oligocene layers; two of these contain great quantities of fossil wood and abound in fossils. On top is a layer of basalt that has protected the hill from erosion. *Top right:* a cranium of Aegyptopithecus, the biggest and best known of the Fayum apes.

said to occupy an intermediate position between Omomyidae and the narrow-noses; like the former, they have three premolars in each half of the jaw, but their dental structure is otherwise more like that of the narrow-noses. One may assume that the known Parapithecidae make up a collateral branch (from our point of view), but that present-day narrow-noses probably passed through a Par-

apithecidae stage when the Eocene period was succeeded by the Oligocene. Remains of such a conceivable transitory type are known through finds in Burma from the early Eocene period. Unfortunately these remains are quite fragmentary.

Aside from some incompletely known types, the other main group is made up of three or four varieties, which have led to the two genera *Propliopithecus* and *Aegyptopithecus*. (Some scientists believe that their differences are so small that they ought to belong to the same genus, which would mean keeping the older of the two names, *Propliopithecus*.) They belong without a doubt to the superfamily Hominoidea; in other words, they are the oldest known members of this entire group. The number of their premolars in each half of the jaw has been reduced to two. They thus have thirty-two teeth in all, just like narrow-noses of the present day. Furthermore, they have stubby molars with separate cusps, like those of the Hominoidea but unlike those of the other current group of narrow-noses, namely the tailed monkeys (macaques, bavians, and many others), the superfamily Cercopithecoidea. So far everyone agrees. But viewpoints differ as to which role

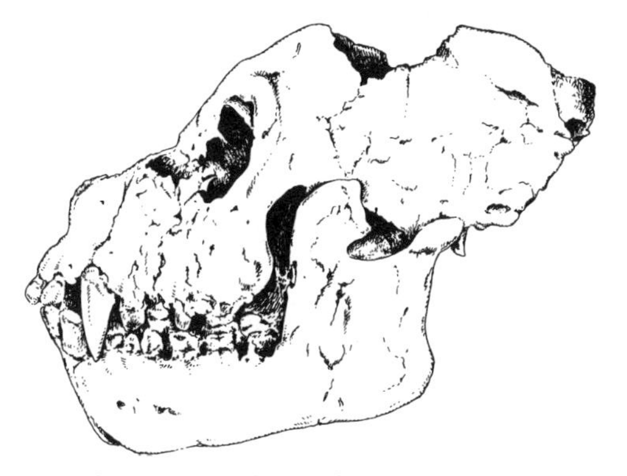

Figure 4.3 The earliest skull of Aegyptopithecus found at Fayum. Several specimens are now known, and they show that the males were bigger and heavier than the females.

these monkeys have played in developments. One assumption is that they occupy the direct line of development between the great apes and us. This is possible, but it would be an unexpected piece of luck if this tiny glimpse of Africa's Oligocene monkeys happens to throw light on our own ancestors. Another view has it that the Fayum Hominoidea belonged to a now-extinct group, Pliopithecidae, which gave rise to the Miocene *Pliopithecus*.

In spite of this uncertainty, these anthropoid apes from Fayum give us an idea of the general characteristics of the earliest hominoids. The best known is *Aegyptopithecus zeuxis*, a creature with a head the size of a cat's. Its dental structure reminds us of that of the great apes, with strong canines and a cone-shaped front premolar in the lower jaw, which with the canines creates a sharp cutting mechanism. The males were noticeably bigger than the females and had more powerful canines. Even the molars were of the same type as those of the great apes. On the other hand, the head is otherwise archaic in its anatomical details as well as in the size of the brain.

The relationship between brain size and body size is not as simple as one might think; we can illustrate this through a comparison with the human brain. We could point out that the male on average has a larger brain than the female. But doesn't that depend on the fact that the male on the whole is taller and bigger? How does it work out if we instead take the relative volume—the brain as representing a percentage of the body volume? In this case the female on average has a larger brain than the male.

Actually, in each case there is a miscalculation. The brains of men and women are exactly equivalent, and the relationship between brain size and body dimensions follows an exponential curve of the type

$$\text{Volume of the brain} = a \text{ times (body size)}^{2/3},$$

which implies that if body size triples, brain volume doubles. Thus we see that the brain grows more slowly than the body. (However, the value 2/3 is not exact; the real value approximates 0.6). Naturally, not all the values are precisely in line—individual variation is great, and the equation merely describes the general tendency.

Thus, the bigger the body, the smaller the brain in relation to the body (in spite of increasing actual body size). The opposite is also true: the smaller the body, the larger the brain in relation to the body. (In a child, the head is much larger in relation to the body than in a fully grown person.) This phenomenon is called allometry—a different pace of growth displayed by different body parts.

Aegyptopithecus was a tiny monkey. If it had had a brain

equivalent to, for example, that of a modern chimpanzee, its braincasc would have been enormous; instead it was rather modest in relation to the entire body. If we could imagine that its braincase could have grown to the size of the brain of a chimpanzee, the brain would have been much smaller than that of the chimpanzee, perhaps only half the size.

Only a few bones are known that were part of the *Aegyptopithecus* skeleton (aside from the skull and the lower jaw). An analysis of these fossils indicates that *Aegyptopithecus* was a very good climber who moved easily on all fours along tree branches and even on the ground. An original characteristic is that *Aegyptopithecus,* like its Miocene successor *Pliopithecus,* had a short tail. All present-day hominoids lack a tail.

The other Fayum monkey, *Propliopithecus,* is smaller than *Aegyptopithecus* and is not as well known; the finds are limited to jaws and teeth. They are rather like those of *Aegyptopithecus,* even though the canines are less well developed. Two or three different kinds are known. At any rate, with *Propliopithecus* we seem close to the earliest hominoid—the common ancestor of all Hominoidea.

Here we reach the end of the records of the Oligocene period, and a gap appears in fossil history. Only in recent years has it become clear that the gap comprises nine or ten million years. And we have no fossil documentation of this tremendous length of time at all. Thus there is still much to be discovered in Africa.

CHAPTER FIVE

"Miocene Lady"

Miocene Lady was the name of the venerable motorboat with which Louis and Mary Leakey and their collaborators crossed over to Rusinga Island in Lake Victoria, near the border between Uganda and Kenya. Located here is the western flank of the Miocene basin, which twenty-three to fourteen million years ago was filled with fossil-carrying river and lake sediments. It comprises southwestern Kenya and eastern Uganda. By the 1930s Rusinga Island had become well known for its rich finds of Miocene apes. Many other discoveries have been made since then, especially in Kenya.

The Miocene epoch began twenty-five million years ago and ended five million years ago with the transition

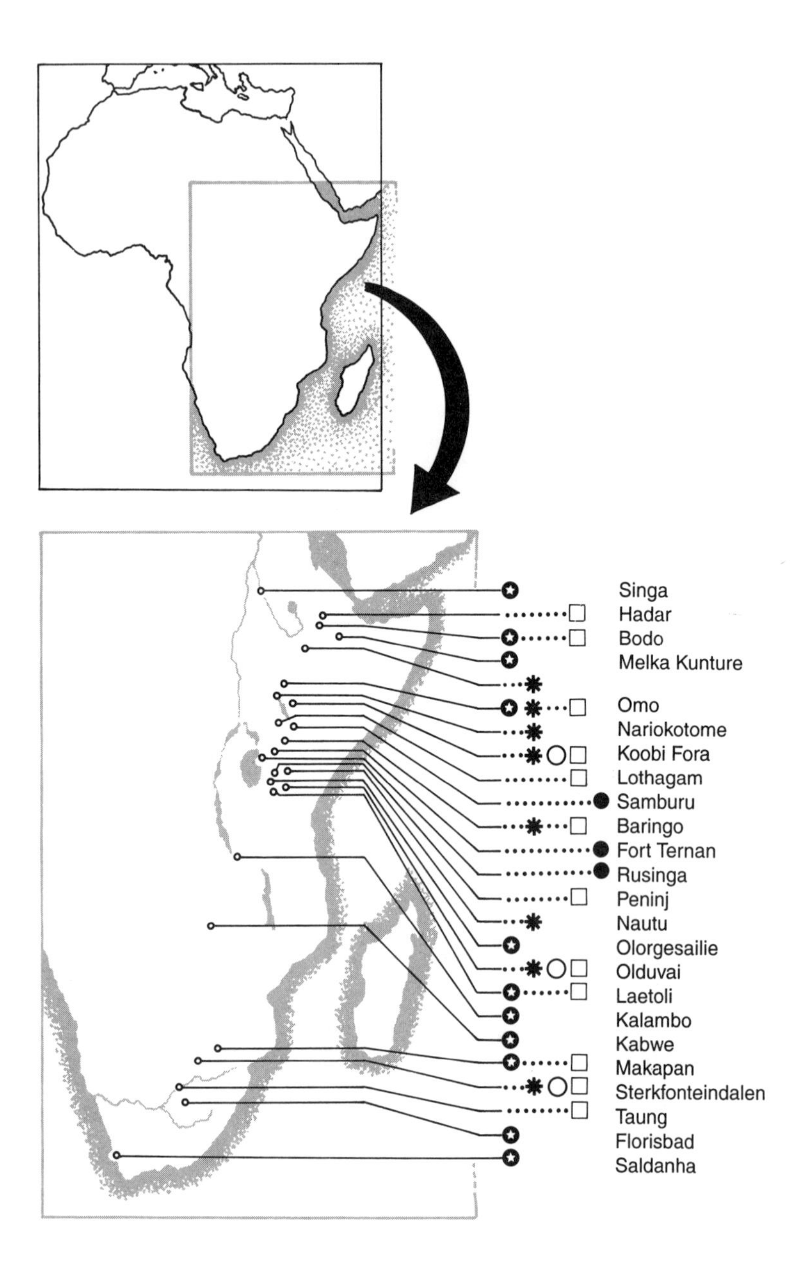

Singa
Hadar
Bodo
Melka Kunture
Omo
Nariokotome
Koobi Fora
Lothagam
Samburu
Baringo
Fort Ternan
Rusinga
Peninj
Nautu
Olorgesailie
Olduvai
Laetoli
Kalambo
Kabwe
Makapan
Sterkfonteindalen
Taung
Florisbad
Saldanha

to the Pliocene period. Most of the places in Africa where apes have been discovered date from the middle Miocene (eighteen to twelve million years ago); the Rusinga apes are about eighteen million years old. The same is true of finds on the neighboring island of Mwfangano and in a number of localities on the mainland in Kenya (Loperot Songhor, Koru) and in Uganda (Napak, Moroto), their ages varying between approximately seventeen and twenty million years. Finds from a few localities are somewhat younger (Maboko and Fort Ternan, fifteen and fourteen million years old, respectively), and a few somewhat older (Bukwa and Karungu, twenty-three and twenty-two million years old, respectively). Thus we have a gap of nine to ten million years between the Fayum apes and the oldest Miocene finds, and unfortunately an equally long void also exists in the upper Miocene period. But the material now available from seventeen to twenty-three million years ago in Kenya and Uganda is respectable: more than a thousand examples of higher primates.

Certain conclusions can also be drawn about the landscape. It seems to have been varied, for the most part forested but with glades and open grassy fields. The rich fauna indicates that Africa had become less isolated. However, in the older Miocene period an arm of the sea still extended eastward from the Mediterranean basin to the Persian Gulf and the Indian Ocean, constituting a barrier

---

Figure 5.1 Localities in Africa where fossil primates have been found. The diagram shows the main areas in eastern and southern Africa where finds have been made; the outline map (*below*) indicates localities where hominids as well as apes have been found.

for any exchange of fauna between Africa and Eurasia. The hominoid primates were still endemic to Africa.

The largest animals were mastodons, primitive relations of our elephant, with blunt and knobby molars and tusks in both the upper and lower jaws, and *Deinotherium,* an animal similar to the elephant and having downward-directed tusks in its lower jaw and none at all in the upper jaw. The hyraxes, relatives of the elephant animals, are represented as at Fayum by large-size specimens. Both these groups, like the hominoids, were endemic in the older Miocene period. Despite that, proof exists that a path connected Eurasia and Africa during the transition between the Oligocene and the Miocene periods. Even the early faunas of the Miocene period show that a massive immigration took place, including insectivores, bats, rodents, swine, giraffes, antelopes, rhinoceroses, and other hoofed animals, in addition to carnivores belonging to the dog, cat, and civet cat families. Thus, the animal world completely changed its character following the Oligocene period and became dominated by "cosmopolitan" mammals.

Much has happened to the apes since we lost their trail in the middle Oligocene period. The primitive Parapithecidae with three premolars in each half of the jaw are gone. In their place we now find three large groups of Catarrhini apes—that is, narrow-noses. One is the genus *Pliopithecus,* which seems to have been a direct continuation of the Oligocene *Propliopithecus* and *Aegyptopithecus.* They were small, slimly built anthropoid apes with long arms and also rather long legs. They show some similarities with the present-day gibbons; however, they were

probably not the ancestors of today's gibbons. From our point of view, the most interesting group is the great apes, but before we discuss them we should point out that the third group comprises the earliest members of Cercopithecoidea, that is, the superfamily that includes present-day macaques, baboons, and patas monkeys. Strangely enough, it looks as if Cercopithecoidea evolved from the earlier hominoids and not the opposite, which had once been assumed. The oldest Cercopithecoidea that we know of appeared in Africa eighteen to nineteen million years ago. However, they did not actually flourish until the late Miocene and Pliocene periods.

We are now getting to the great apes and will for the sake of simplicity call them pongids (the family Pongidae, which also includes the great apes of the present day, but not the gibbons), despite the fact that this is a paraphyletic group, as indicated in chapter 1. Most of the Miocene pongids in Africa may be traced back to the same subfamily, Dryopithecinae, named after the genus *Dryopithecus* ("tree ape"). Best known is *Proconsul africanus,* of which we have made numerous finds, including large parts of the skeleton. It appears that this ape, which was about as big as a baboon, moved on all fours. Arms and legs were of about equal length, so the spine was horizontal when the animal was standing on all fours; the pongids of today that move on all fours assume a half-upright posture, since their arms are much longer than their legs. *Proconsul*'s skeletal anatomy indicates that, like the Oligocene *Aegyptopithecus,* it lived in trees and ran across branches, which it grasped with its hands and its handlike feet. This is quite different from the way today's great apes move; they prefer to hang by their arms from the tree branches in what is called *brachiation*. Judging from skeletal finds, brachiation had not yet developed among the apes we know from early and middle Miocene.

Proconsul's cranium shows that its development had advanced considerably from the *Aegyptopithecus* stage. The brain is bigger relative to the face, even through *Proconsul africanus* is a bigger type than *Aegyptopithecus zeuxis* and thus ought to have had a smaller braincase, relatively speaking. On the other hand, *Proconsul* had a considerably smaller braincase than do current great apes of the same size. Its brain size is midway between that of Oligocene apes and today's great apes.

From the structure of the teeth and the way they have been worn down, we can conclude that *Proconsul* lived mainly on fruits.

The males were considerably bigger than the females, which is also the case among many other primates. This may indicate a social organization with flocks led by a dominant male.

Two larger types of the same family are also known, mainly on the basis of jaws and teeth. One of them, *Proconsul nyanzae,* was about as big as the present-day chimpanzee, while the other, *Proconsul major,* was the same size as a gorilla. Of the latter, parts of the face have

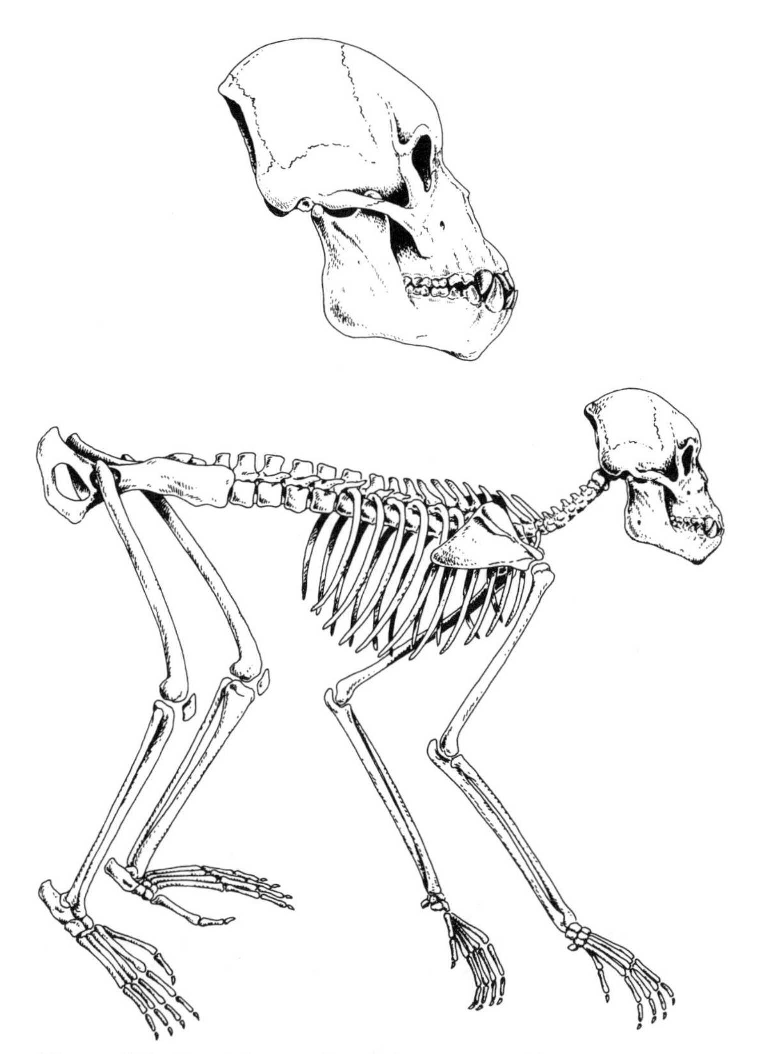

Figure 5.2 *Top:* The cranium of Proconsul, the best known of Africa's fossil anthropoid apes. *Bottom:* The skeleton of Proconsul shows a powerfully built, four-legged anthropoid ape with no ability to hang by the arms or walk upright.

been found. We know several other pongid families, and we will return to them later.

Seventeen million years ago, an important geographical change occurred. The ocean bay that had long isolated the African-Arabian lowlands from Eurasia finally dried and an intensive exchange of fauna between Africa and Eurasia began. Hominoids as well as cercopithecoids emigrated to Europe and Asia; even elephants and hyraxes wandered north. The hyraxes never became very successful, while the elephants started a tremendous march of conquest that in time led them to North and South America as well as Europe and Asia. The geographical changes are connected with the movements in the Earth's crust that were the immediate cause of the rise of young "alpine" mountain ranges in Europe, Asia Minor, and Iran.

Different species of *Dryopithecus* lived in Europe in the Miocene period between sixteen and eight million years ago. Several types were probably represented; there is much disagreement as to the systematics. As far as is known, the European Oryopithecus was not the direct ancestor of either the human race or any primate currently living. We go on to the Asiatic *Dryopithecus,* which most probably included the direct ancestors of the orangutans.

The most important finds in Asia took place in the Siwalik Mountains in northern Pakistan and India. They represent three genera: *Sivapithecus, Ramapithecus,* and *Gigantopithecus.* The same genera also appear in China, and at least two of the first-named lived in Asia Minor and possibly in Eastern Europe.

Sivapithecus (pronounced shiva-), the best known, lived in Eurasia fifteen to eight million years ago. It was an anthropoid ape the size of an orangutan and is thought to be closely related to and perhaps even the direct ancestor of this big anthropoid ape, the only one of its kind living in present-day Asia. A well-preserved face of *Sivapithecus* indicates far-reaching similarities with the face of the orangutan. Certain details in the form of the palate that separate the orangutan from chimpanzees, gorillas, and human beings are, for instance, of the orangutan type in *Sivapithecus.* Among other things, it differs from the orangutan in that its teeth have a thicker layer of enamel. This feature, as well as several others, it shares with the genera *Ramapithecus* and *Gigantopithecus.*

Ramapithecus, a contemporary of *Sivapithecus,* was considerably smaller, about the same size as the present-day dwarf chimpanzee. With its short nose, its thick layer of tooth enamel, its fine row of teeth without any gaps, and its relatively small incisors and canines, it looks conspicuously humanlike. It had long been regarded as a genuine hominid and a direct precursor of the human types. However, it shows such great resemblance to the better-known

Sivapithecus that most experts now believe that the two of them are closely related to each other and to the orangutan rather than to the human.

The same seems to be the case with *Gigantopithecus,* a giant anthropoid ape bigger than the gorilla—and the biggest primate ever known. Of the genus *Giganthopithecus* we know two types, one from the Siwalik area (eight to ten million years old) and a much later (and bigger) type from China, which may still have been living a million years ago or later (quite often one sees references to *Gigantopithecus* in connection with the mythic "abominable snowmen" or yeti). Off and on, even *Gigantopithecus* has been regarded as a possible ancestor of the human race; the similarities are basically the same as with *Ramapithecus*. Today, all three types are regarded as belonging to the orangutan family.

We have already learned that our closest relatives are the African anthropoid apes, the chimpanzees, and the gorillas. It is therefore probable that the hominids originated in Africa. On the other hand, a separate orangutan lineage evidently began in Asia sixteen million years ago. This provides us with a probable minimum age for the branching at point B in figure 1.2. Nothing prevents us from assuming that the branching took place even farther back in time, as, for instance, seventeen million years ago, when the straits between Arabia and Southwest Asia became dry. It may have taken place even earlier: the *Sivapithecus* group and the chimpanzee-gorilla group may have had separate ancestors even in the early Miocene period.

Recently a *Sivapithecus* was discovered in East Africa, in strata located at Buluk that are somewhat older than seventeen million years. This would indicate that the genus developed in the transition period between early and middle Miocene in Africa and later emigrated to Eurasia. Some scientists believe that this early African *Sivapithecus* was also a predecessor of the African chimpanzee-gorilla-human being line. But was the find actually a *Sivapithecus?* According to another, perhaps more likely, interpretation, we are dealing here with the family *Kenyapithecus,* to which we shall return later.

Given the many African Pongidae of the Miocene period, it is difficult to select the ancestral line that led to the hominids. The problem is compounded because remains, apart from teeth and jaws, are so negligible. *Kenyapithecus,* which was found in Kenya (fourteen million years old), shows a strong resemblance to *Ramapithecus,* which implies that it has humanlike traits. Perhaps it belongs to the same family that appears in Eastern Europe in the later Miocene period (twelve to six million years ago), although these types also have led back to *Ramapithecus.* Once again it is a question of very small anthropoid apes, the same size as dwarf chimpanzees or six-year-old children, with relatively small canines and thick tooth enamel.

It is completely possible that *Kenyapithecus* belongs somewhere between points B and C in figure 1.2. However, it is impossible to make a final determination, since the trail stops here. After the finds at Fort Ternan, which are fourteen million years old, a gap appears in the history of African primates, which except for a few fragmentary finds extends until 3.7 million years ago.

The chances look good that the gap will be filled. For example, at Lake Baringo in Kenya an extensive series of strata exists, with many fossils that may be dated to the early to middle Miocene period. Other promising series of layers have also been found there.

Another kind of gap exists that is perhaps more serious. We do not have a major part of the skeleton of any type besides *Proconsul africanus*. The great surprises it provided scientists show how important it is to gain insight into the entire anatomy rather than just the cranium, the teeth, and the jaws, which traditionally have been the most important objects for study. David Pilbeam has said about this skeleton that the elbow and shoulder joints resemble those of the chimpanzee, the wrists are like those of the cercopithecid, and the lower vertebrae resemble most closely those of the gibbon, while many other characteristics are unique, as is the combination of them all. Even if, like Pilbeam, one talks of a "mosaic type," it might as well be said that each variety may be regarded as a mosaic—one

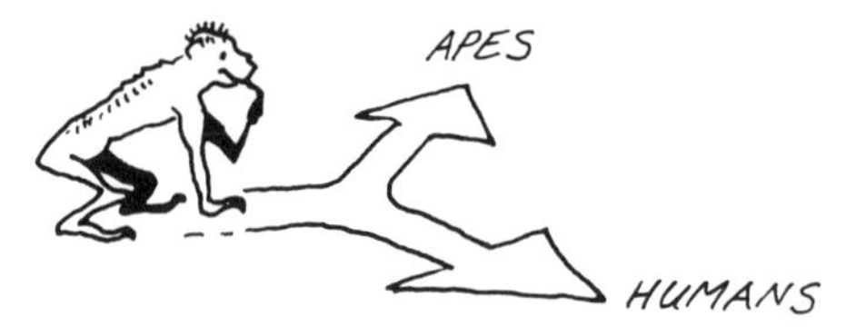

may also say that, for example, the gibbon shows a part of the mosaic of *Proconsul*.

The ongoing research into the phase of development linking us with the apes may thus bring with it many exciting surprises.

CHAPTER SIX

The Fork in the Road

As this is being written (1986), human prehistory in Africa is practically devoid of fossils from the period between fourteen million and four million years ago. Except for one tooth or another, the only find has been a jaw about nine million years old, and so far it has not been scientifically described. But as soon as we go beyond the four-million-year mark, we gradually meet with almost overwhelmingly rich documentation.

All the earliest known hominids (in other words, members of the zoological family of humans, Hominidae) belong to the genus *Australopithecus* (the name means "southern ape"). We have no definite finds of *Australopithecus* other than from Africa south of the Sahara, so it

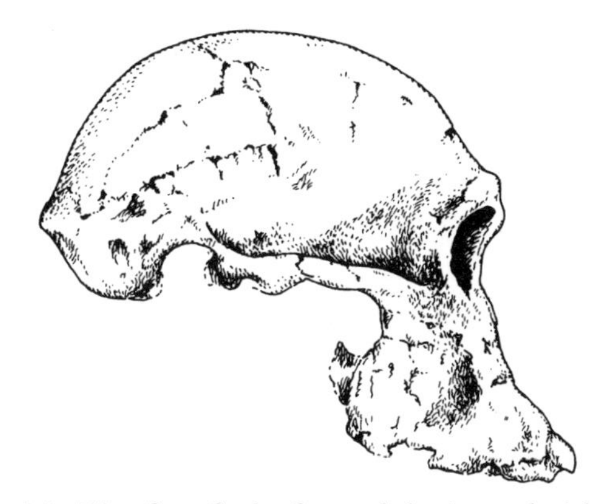

Figure 6.1 The first find of an adult *Australopithecus* was this skull from the Sterkfontein grotto in South Africa. The teeth, however, have been lost. The skull has been given the pet name "Mrs. Ples."

appears that these hominids were endemic to that region. The finds were made in two separate areas: South Africa and East Africa (Ethiopia, Kenya, Tanzania).

The first discoveries occurred in South Africa, starting in 1924. They were all made in sediments in ancient caves, so it has been difficult to arrive at accurate dates—radiometric methods are not suitable to dating sediments within caves. When finds were made in the great rift valley in East Africa, the picture radically changed. Over millions of years this valley has been the destination of sedimentary material from the surrounding, more mountainous areas. In these sediments, which unlike the South African ones were deposited under the open sky, layers of volcanic ash often exist that have been assigned fairly exact dates. Over millions of years, the volcanic activity in the rift valley was intensive, and the repeated eruptions resulted in a detailed series of date lines in the layers. The sediments abound in fossils, not only of hominids. By correlating them with corresponding fossils in South African caves, we can determine the exact ages of the latter. Thus we have a well-substantiated chronology for these protohumans.

Australopithecus walked upright in the same manner as modern humans. This is plainly shown by anatomy, as, for instance, by the structure of the pelvis and the knee joints. The human, in contradistinction to the anthropoid apes, is knock-kneed: the thigh bones converge toward the knees, while the shin bones are straight and vertical. In this way, the foot is close to the vertical line from the center of gravity. The apes' walk may be likened to a "jog trot," since their center of gravity must be shifted from side to side for every step. Further, *Australopithecus,* like humans, had an S-shaped spine. Neck joints are under the cranium, which shows that the head rested on a vertical neck. All this was dramatically verified through the discovery of fossil footprints at Laetoli in Tanzania.

The footprints were made in volcanic tuff originating from the nearby volcano Sadiman. They were made during the rainy season, imprinted in wet ash mud that later dried and became hard as rock. Such occurrences have been repeated several times, since we find several different layers of footprints on top of one another. Most of the footprints have been made by different animals, but many of them were made by *Australopithecus,* and they show that their way of walking was just like ours. By 3.7 million years ago—the age of the Laetoli sediments—the hominids were accomplished two-legged creatures.

Their body lengths vary from approximately 1.1 meters

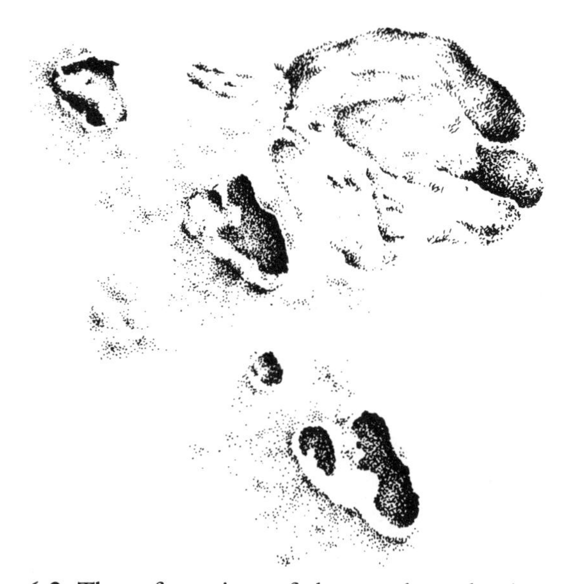

Figure 6.2 These footprints of the two-legged primate *Australopithecus afarensis* were made 3.7 million years ago in ash from the volcano Sadiman in Tanzania. The tracks have been preserved intact in the hardened ash. Next to the footprints of *Australopithecus* are animal tracks (among them an elephant's).

to perhaps 1.8 meters; generally it increases over time. Males were considerably bigger than females. In one series of footprints, two members of *Australopithecus* have been walking together—the lengths of their bodies, which may be estimated by the length of their paces and from the size of their feet, were approximately 1.4 and 1.2 meters, respectively. At one point the smaller of the two individuals stopped and momentarily turned left.

While *Australopithecus* had a typically "human" body, its head was very much unlike that of the modern human. The braincase was on average approximately 450 cubic centimeters in volume, that is, about one-third of the 1,350 cubic centimeters of the brain of the modern human. Brain size tended to increase noticeably over time, but this may have been merely a direct consequence of increasing body size. The brains of the great apes are about the same size as that of *Australopithecus,* but the great apes are much heavier; an average *Australopithecus* weighed between twenty and forty kilos, while a gorilla weighs more than one hundred kilos. The brain of *Australopithecus* was thus an advanced one, compared with that of present-day anthropoid apes.

Dominant in the face of *Australopithecus* are the powerful jaws with their very large teeth. The teeth are similar in shape to those of human beings, but the molars are enormous; the canines are small and pointed but through usage assumed the form of chisels; the incisors are markedly small compared with those of the anthropoid apes. It is evident that the battery of molars played the main role, while the incisors and canines played a secondary role.

Thus *Australopithecus,* compared with its predecessors, is marked by an important innovation: walking upright. Other important innovations are the enlargement of the molars and the somewhat more highly developed brain.

In the following discussion we will place the chimpanzee and the gorilla together in the genus *Pan* (alternatively, one may place the chimpanzee in *Pan* and the gorilla in the genus *Gorilla*). The development of the upright position must have occurred after point C in figure 1.2, after the hominids separated from *Pan.* In the opposite case, we would be able to trace some of the

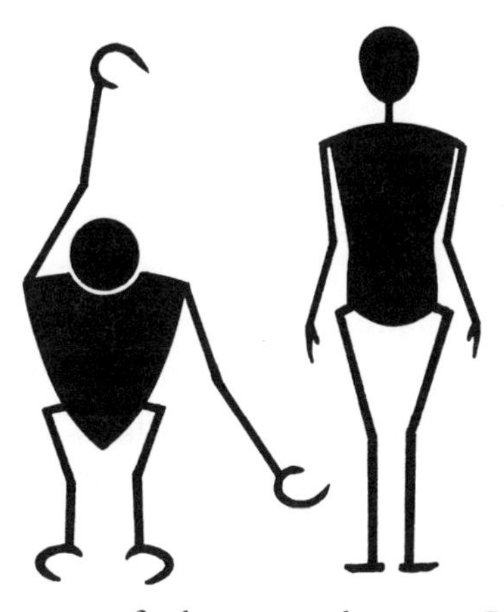

Figure 6.3 Postures of a human and an ape. Because humans are knock-kneed, the feet are located almost exactly below the center of gravity. An ape that walks on two legs must shift the body sideways for every step and thus displays a waddling walk. Instead, the arms of the ape are "knock-kneed," so that when it hangs by the arms the hands will be right above the center of gravity. Even in human beings (especially among women) the forearm is bent at an angle at the elbow so that it doesn't brush against the hip.

complex anatomical characteristics associated with this change, even in *Pan*. But there is no trace of them. There are no indications that *Pan* ever went through a stage characterized by an upright walk, then went back to movement on four legs. The specialization in *Pan* is completely different: a way of life characterized by hanging by the arms and by short legs and long arms (which are knock-kneed in precisely the same fashion as the legs of the hominids).

Thus, after the branching at C the hominids and *Pan*

took completely different routes of adaptation. *Pan,* like its cousins the orangutans and the gibbons, adapted themselves to hanging by the arms, while the hominids became two-legged animals on the flat ground. If it is correct that C, the last progenitor they had in common, lived seven to nine million years ago, the upright stance must have been assumed between that time and approximately four million years ago, when we encounter *Australopithecus* as a full-fledged two-legged creature.

Why did humans get up on two legs? How did it happen? These two questions have been discussed ever since the time of Darwin. It is impossible here to discuss the innumerable scenarios that have been proposed. I speak intentionally about scenarios, not theories or hypotheses, since the available facts make for a meager assortment.

It is often assumed that our ancestors had adopted a way of life characterized by hanging from branches by the arms, which provided the first incentive toward assuming an upright position, but that they had abandoned this within a short time in favor of getting down on the ground before their arms grew longer and their legs grew shorter. But although it may seem probable, there is no proof of this contention, and the same is true of practically all the scenarios.

Our ancestors seem to have adopted a two-legged existence on the ground at the same time that the baboons became four-legged creatures on the ground. One can assume that for our ancestors a two-legged life on the ground implied an important advantage. It cannot have had anything to do with speed in running, since a four-legged creature runs much faster than a two-legged one. It might have had something to do with the elevation of the head above the ground, that is, gaining a further, more comprehensive view—which, however, doesn't sound as if it is a decisive advantagc. Most probably it had to do with the function of the hands, since two-leggedness frees the hands to grasp and manipulate. This is a chief theme in most of the scenarios.

In the opinion of some, we became human when we produced and used tools. This requires free and usable hands. Later, a related scenario was proposed by Friedrich Engels, who believed that *work* gave rise to humanity. These are without a doubt important viewpoints, but they seem a bit too reductionistic to be convincing.

One of the most recent scenarios, created by Owen Lovejoy, focuses on the ability to *carry with one* various objects, made possible by leaving the hands free. In this case it has to do with *food*. Among the higher primates the young are wholly dependent on their mothers during the first four to five years. The mother seeking food is forced to take her young along, which makes it impossible for her to have a big family. If on the other hand she can build a dwelling where she can stay with her small ones, the male can be charged with looking for food—provided he is able to carry it home with him. Here we have a

prototype of the nuclear family, which in this case implies that the female can give birth to several babies in quick succession, thus obtaining a selective advantage.

Lovejoy's scenario is brilliantly thought out, but it is of course impossible to prove.

Another scenario, submitted by Alister Hardy, is dealt with by Karl Erik Fichtelius in the book whose translated title is *How the Ape Lost Its Fur Coat*. It maintains that our ancestors—but not *Pan*'s—once lived as swimming denizens of coastal areas, among other things seeking food on the bottom of the sea (mussels and other shellfish are very nourishing). Submerged in water, one can assume any position, as, for instance, the upright one. What is less clear is why an upright position in the water would lead to an upright stance on dry land.

On the other hand, rather convincing proof of an amphibian stage in our development does indeed exist in our anatomy, and perhaps in our behavior.

Hairless skin is a feature we share with many marine mammals (whales, seals, and hippopotamuses). Hairlessness, however, is also found in animals living on land (elephants), and in certain cases among burrowing animals (sightless rats). However, human beings (especially

women) also have a well-developed layer of fat underneath the skin, which creates a streamlined appearance. In the animal kingdom, the combination of hairlessness and subcutaneous fat is known only among marine mammals.

It is difficult to think of any other convincing explanation for this characteristic. Attempts to explain hairlessness have included the need to perspire unchecked in a tropical climate. (If so, the subcutaneous fat seems to have some other reason for its existence.) An alternative explanation for the combination of hairlessness and subcutaneous fat is that sexual attraction and intimacy are thereby increased. The human is a highly sexual being, since the female can mate independently of her monthly ovulation.

In the "amphibian theory" the hair on the scalp is said to have provided swimming children an opportunity to cling to the mother. Scalp hair has also been thought of as acting like a sun visor, etc. The male beard has among other things been regarded as a sexual and social status symbol, and the same is true to a certain extent of hair around the genitalia. On the contrary, hair in the armpits

Figure 6.4 At one stage of the development of the human fetus, it is covered by a thin layer of hair. The direction of the hair corresponds to the movement of water along a swimming body.

and around the anus might have developed to spread odors produced by the proximate glands.

At one stage the human fetus displays hair over areas of the skin, but it soon disappears. The direction that this hair assumes corresponds to the direction in which water flows across one's skin during swimming. The skin of our ancestors, when they changed over to life in the water, might have been covered with thick furlike hair that at the beginning assumed a direction that facilitated their swimming. Later on, when the heat-isolating subcutaneous fat developed, the fur could be dispensed with.

Another anatomical peculiarity that may indicate a past life in the water is the existence of the hymen (the maidenhead) in human females, just as in whales and seals. The hymen is not found among apes. Its purpose may have been to prevent water from entering the vagina. The sharp bend in the human vagina may have had the same function; it is much less pronounced or is completely missing among the *Pan*. To this we may add the great ability of human beings to swim and dive, which far surpasses that of the apes (even though swimming among human infants has a counterpart among the anthropoid apes).

Thus, the scenario of our possible amphibian past is based on a great deal of probability and can above all explain the combination of hairlessness and our subcutaneous fat, which is lacking among the apes; as the sole explanation of our two-leggedness, however, the scenario seems unsatisfactory. Combined with certain other processes (freeing the hands to carry and manipulate objects) it may sound more believable—even more so if we include the fact, pointed out by Lovejoy and by Desmond Morris, that the human is a sexual being to a much higher degree than any ape. We must in that case look for this amphibian stage somewhere between seven to nine million years ago (point C of figure 1.2) and a time approximately four million years later that marks the appearance of the first-known *Australopithecus*.

Our body is in a way a historical museum showing our ongoing development and can provide us with many pointers regarding our origin. But to learn what *actually* happened we need fossil documentation. This doesn't mean that speculations like those of Engels, Lovejoy, Hardy, and others are worthless. They show that there may be rational explanations of events that may seem undecipherable, and in certain cases they may point to promising areas for research.

If we assume that an upright walk and hairlessness developed at about the same time and in a more or less reciprocal manner, we have received help in the reconstruction of *Australopithecus*. The bones that have been

unearthed make it possible for us, in a broad outline, to get an idea of the appearance of these creatures. The hypothesis about their amphibian past indicates that *Australopithecus* probably had skin as hairless as that of people today.

CHAPTER SEVEN

Hominids Before Homo

In the last chapter we glanced at the genus *Australopithecus* as a whole. However, we do know that there have existed several species of this genus: at least two, perhaps as many as four, species may be differentiated on the basis of the material that has become available.

One usually talks about two main types—the slender or gracile and the robust *Australopithecus.* All the earliest ones are the slender type, while all the later ones are the robust type. But it does appear as if both types existed at the same time.

The oldest ones are often assigned to a species of their own, *Australopithecus afarensis,* whose remains were found at Laetoli in Tanzania (and which was the animal that

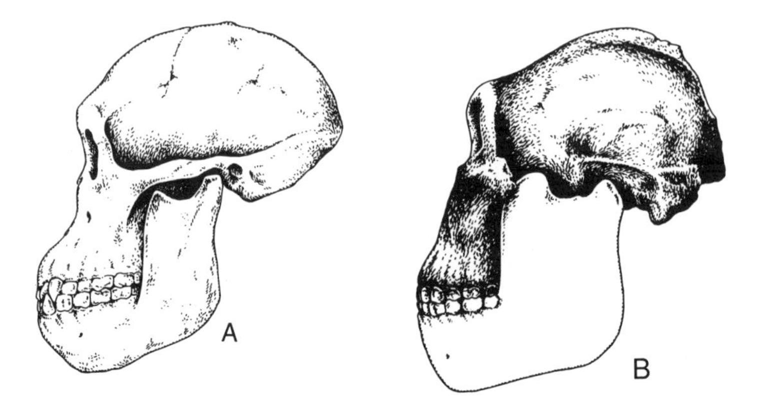

Figure 7.1 Gracile (*A*) and robust (*B*), Australopithecus. (*B*) the younger one, is assumed to have descended from (*A*). The development shows a reduction in the size of the incisors and canines as well as an enlargement of the molars and related changes in the shape of the skull and lower jaw.

made the footprints) and also at Afar in Ethiopia. Thus it is not known to have lived in South Africa. The finds vary in age from almost four million years to approximately three million years. Like the other *Australopithecus, A. afarensis* had a low forehead, a flat nose, protruding jaws, broad cheekbones, and no protuberant chin. Its braincase was between 370 and 500 cubic centimeters in volume. Although *A. afarensis* could walk just like a present-day human being, its slightly bent finger and toe joints may indicate that it was also an able tree climber.

Australopithecus africanus is the other slender or gracile type. It was discovered in 1924 and was described in 1925 by Raymond A. Dart from Johannesburg as being an intermediate form between apes and human beings. The find, made in Taung Cave, consists of the skull and lower jaw of a child with milk teeth and a molar in each half of the jaw. Dart's opinion was met with skepticism at first but has since been generally accepted.

Other finds in South Africa took place in Sterkfontein Cave and Härd Cave at Makapansgat. Even in East Africa there have been finds related to this species—at Omo in Ethiopia, Koobi Fora in Kenya, and Olduvai in Tanzania. Remains from Olduvai are somewhat larger than *A. afarensis,* with braincases of 425 to 485 cubic centimeters (440 on average). In other aspects the similarity is great, and some specialists think that *A. afarensis* at its acme was a subspecies of *A. africanus.* The borderline between the two is to a certain extent vague, as is also the case with *A. africanus* and the robust types, which will be discussed below.

Australopithecus africanus seems to be the more common type during the two to three million years of its existence, but the slender *Australopithecus* is found in somewhat younger sediments (as, for example, at Olduvai, where the entire sequence of layers is less than two million years old).

Like the slender ones, the robust *Australopithecus* was first discovered in South Africa. These finds, made by Robert Broom, were made in the caves of Swartkrans and Kromdraai in the Sterkfontein valley, very close to Sterkfontein Cave with its slender type. However, the caves are not the same age, a fact established early on the basis of many animal fossils found there. Of the three caves, Sterkfontein is the oldest and Kromdraai the youngest.

The South African species has been given the name *Australopithecus robustus.* Most of the finds were made at

Swartkrans. The species is larger than *A. africanus* and has proportionally larger molars, extremely powerful jaws, and only a slightly larger brain (530 cubic centimeters in the only cranium that has been able to be measured). The muscles for chewing were well developed. The zygomatic arch of the skull is very wide and thick and indicates that the masseter muscle, which runs from the arch to the lower jaw, was powerful. A large temporalis muscle is indicated by the presence of a bony ridge on the side of the skull where that muscle attached. The age of *Australopithecus robustus* may be set at 1.5 to 2 million years.

The robust species of East Africa, *Australopithecus boisei,* which also has been regarded as a subspecies of *A. robustus,* shows the same characteristics, but to a much higher degree. Many discoveries have been made at Omo, Koobi Fora (and other places by Lake Turkana), Olduvai—where the first discovery was made by Mary and Louis Leakey—as well as other localities. The species seems to have become extinct about 1.2 million years ago, marking the disappearance of the genus *Australopithecus.* The oldest discoveries are more than two million years old. The material is extensive, and new finds are frequently reported. Braincase volume is generally between five hundred and six hundred cubic centimeters. Thus, compared with *africanus* the increase is rather modest and may be explained solely by the correlated increase in size.

The developmental-historical connections between these groups are rather unclear. The most elementary hypothesis holds that we have a developmental series: *A. afarensis—A. africanus—A. robustus* including *A. boisei.* Certain anatomical details, as, for example, impressions of the

blood vessels in the neck region, have been cited as evidence that *A. robustus* (including *A. boisei*) developed directly from *A. afarensis,* while *A. africanus* is a collateral branch closer to *Homo*. Nor is it clear to what extent the time ranges of the various species overlap. In general, the tendency is toward increased body size coupled with a powerful strengthening of the chewing muscles and an increase in the size of the molars. This probably has something to do with the diet. One view has it that the gracile or slender types were omnivorous, while the robust ones were more specialized as vegetarians: vegetable food (unboiled!) is generally more difficult to chew and creates more wear on the teeth than does animal food.

Is it possible then to determine what kind of food *Australopithecus* preferred? A so-called chipping of the teeth has been pointed out, especially among the robust types, which indicates that their food contained a great deal of sandy particles, which one would indeed expect if the food consisted of, for instance, roots and wooden knobs. Also, scanning electron microscopy has shown that the surfaces of worn teeth take on characteristic wear marks depending on the different kinds of food that have been chewed. The surface wear of the specimens that have been investigated is very much like that found in fruit-eating apes. The look of the worn surfaces was of course mostly influenced by the food intake immediately before the animal's death, though this may not have been a typical meal. Still, it is reasonable to surmise that fruits played an important role in the diet of *Australopithecus*.

An interesting difference between the gracile and the robust types has been observed in the different ages of the slender types at Sterkfontein and the robust types at Swartkrans. At Swartkrans half-grown and young animals were the majority, which indicates a low average lifetime; at Sterkfontein it seems that the average lifetime was longer. The comparison is of course valid only as far as these particular finds are concerned, and that brings up the question of how the bones came to be gathered together in that cave.

Such bones may have come from animals that died in the cave during wintering or hibernating, or having sought shelter during a terminal sickness. They may also have met with an accident, as, for example, falling down a hole leading into the cave. Moreover, they may have been killed by predatory animals (birds of prey, human beings) or discovered by carrion eaters and dragged into the cave. When one finds hominids in caves, it is easy to conclude that they actually lived there. But at least at Swartkrans it seems as if the remains were left behind by predatory animals. Certain skulls display holes made by sharp canine teeth, and the size of the holes and the distance between them exactly fit the mouth of the leopard whose remains were also discovered there. Leopards that hunt baboons usually grab them by the neck and then climb into a tree to eat them in peace. If the tree was located near the cave opening, the skull might have fallen down and rolled into the cave. Perhaps it was chiefly young and inexperienced robust *Australopithecus* individuals that became victims of the leopards.

The East African discoveries originate in sediments by rivers and lakeshores that in many cases are incredibly rich in fossils from mammals, birds, reptiles, etc. Most remains are isolated pieces of bones and teeth, but in certain cases entire sections of skeletons have been found, and at least one find, at Afar ("Lucy," discovered by Don Johanson), comprises almost half an entire skeleton.

Did *Australopithecus* make tools? Great quantities of broken animal bones found in the South African caves have been interpreted to be primitive tools; however, experiments with hyenas and other animals that crush the bones of their prey when chewing have shown that the South African bones look about the same, so these "bony tools" probably have been left behind by hyenas (which are well represented among the fossils in these caves). Baboon skulls from Makapansgat that show signs of having been eaten also appear to have been struck by a club of bone from an animal (the lower part of the upper foreleg of the antelope fits exactly into the impression), which may indicate that the gracile *Australopithecus* actually made use of clubs.

Stone tools are completely unknown in the older sediments containing *Australopithecus.* On the other hand, they occur in great quantities in sediments under 2 million years old (possibly the oldest ones known are from Omo, their age being 2.4 million years). They were origi-

nally thought to have been made by *A. boisei.* Fossil remains representing *Homo* were discovered later in the same sediments, so we are inclined to think that all stone tools have been produced by *Homo*. Even if *Australopithecus* did not know how to make stone tools, we can assume that they *made use of* tools, including rocks, sticks, bones, and anything else within reach. So do the apes of today.

To what extent then can *Australopithecus* be thought of as "human"? We make a bad mistake if we consider them from our self-centered viewpoint, as a kind of half or three-quarters human beings, imperfect creatures on the

way to the human state. On the contrary; like other animals, they were sufficient unto themselves—living in their own world, to which they were well enough adjusted that they got along in it for several million years.

Could *Australopithecus* speak? It has proved difficult to determine the exact degree of the primitive human's ability to speak, but it is not wholly impossible. The sounds that make up a language are formed in the cavity that includes the vocal cords and extends upward into the throat—the so-called pharynx—and further into the oral cavity. The human pharynx is much deeper than that of the apes, and it is here above all that the various vowel sounds are formed. The pharynx is bounded by the underside of the cranium, where in humans it makes a sizable inward bend, which contributes to the great height of the pharynx. Among the apes, who have a short pharynx and can not produce any actual sounds that may be called a language, the base of the cranium is rather flat, as is also the case with all other mammals except man. The deep inward bend under the cranium is thus closely related to the ability to speak. In *Australopithecus* the underside of the cranium is identical to that of the apes, so it is improbable that *Australopithecus* could utter anything like a language.

Another characteristic seems to indicate that *Australopithecus* is much more closely related to the apes than to humans. Childhood and adolescence are long among humans; among apes they are considerably shorter, making a chimpanzee fully grown at the age of ten. Generally it has been assumed that the age of a fossil hominid corresponds to that of a human being in the same stage of

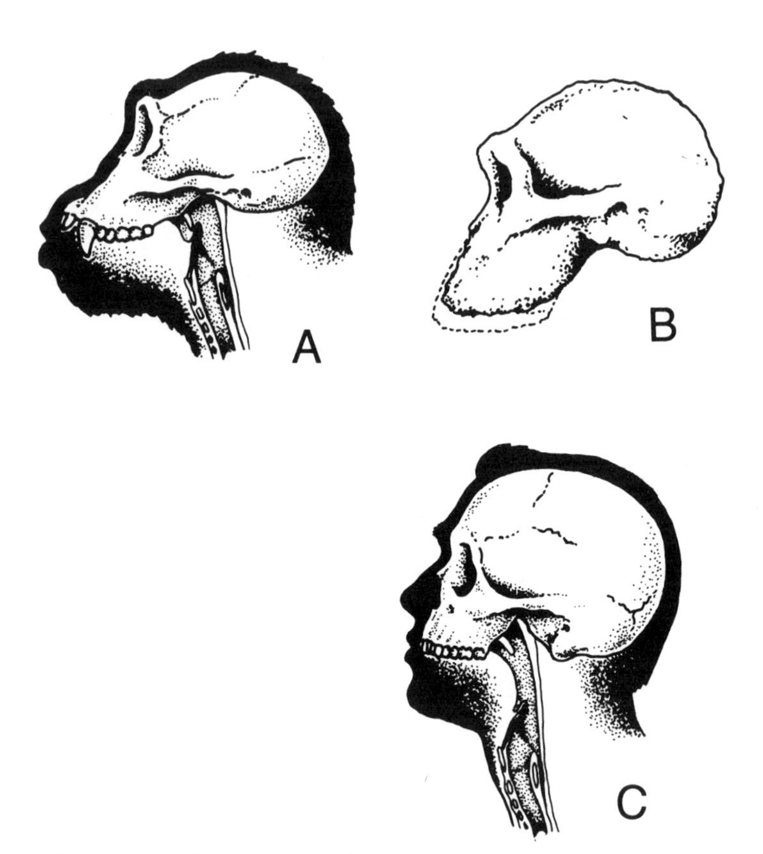

Figure 7.2 Skull and speech organ in the (*A*) chimpanzee and (*C*) human being; between them (*B*) is the skull of *Australopithecus*. In the chimpanzee the vocal cords are located high up, in humans they are much lower; sounds are made in the pharynx. The base of the skull, which serves as a roof for the pharynx, is deeply curved in humans but is almost flat in the apes. The base of the skull in *Australopithecus* most closely resembles that of the ape.

development. Recently, however, scanning electron microscopy of tooth enamel in young members of *Australopithecus* has revealed daily growth lines, which enable us to determine their exact age at the time of death. In all cases their ages have been approximately one-third less than what we would expect in comparison with present-day human beings. Consequently, *Australopithecus* at the age of ten corresponded physically to a modern fifteen-year-old, approximately what is also true among the chimpanzees.

Australopithecus almost certainly comprises the ancestors of our own genus, *Homo*. Among the earliest representatives of *Homo* we find the same characteristics as those of the recently discussed *Australopithecus*—such as a flat underside of the cranium (which indicates an imperfect ability to speak) and a rapid growth rate. The earliest known members of *Homo,* whom we will meet in the next chapter, are approximately two million years old. May some of the known types of *Australopithecus* be considered primitive types of *Homo?*

Without doubt we can exclude the robust types *A. robustus* and *A. boisei*. They had already traveled far on their own road toward specialization, and, besides, they were contemporaries of the earliest *Homo*. On the other hand, both gracile types are conceivable candidates, and both have with full justice been put forward as such. One theory holds that *A. afarensis* gave rise to the robust types, while *A. africanus,* about 2.5 million years ago and as a collateral branch, brought forth the earliest *Homo* (the anatomy of the upper neck may be an indication of that). Other scientists are inclined to date the branching at a much earlier time, before the oldest known *A. afarensis*. Research is intensive, and we feel sure that the matter will be clarified before long.

In its earliest representation, *Australopithecus* was thus in all probability the most primitive incarnation of the human type known as *Homo*. But since *Homo* developed as a collateral branch of *Australopithecus*, that line lived on and developed in its own way, finally dying out about 1.2 million years ago.

CHAPTER EIGHT

Animals and Humans

The earliest positively identified members of the genus *Homo* appear in sediments about two million years old or somewhat older. As far as the structure of the teeth is concerned, *Homo* differs only insignificantly from the gracile *Australopithecus:* the molars are somewhat smaller, the incisors and canines somewhat larger. The differences are so small that one might imagine that *Homo* developed from the gracile *Australopithecus* approximately 2.5 million years ago. This, for instance, is the opinion of Phillip Tobias, while other scholars believe that the origin of humankind occurred one to two million years earlier. It is interesting to learn that the earliest rocks and stones shaped into rudimentary tools have appeared in sediments that

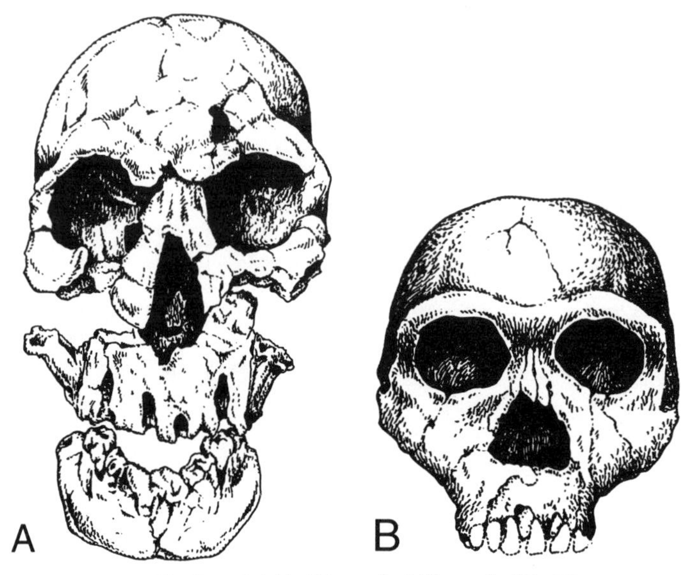

Figure 8.1 Skulls of (*A*) *Homo habilis* and (*B*) *Homo erectus,* both from East Africa.

are 2.4 million years old. Perhaps they were shaped by the earliest representatives of *Homo*.

The oldest known *Homo,* from about 1.6 to 2 million years ago, is usually considered a separate kind, *H. habilis* ("handy man"), known solely through finds in Africa. It is conceivable that older discoveries of jaws and teeth may have belonged to this type, in which case it will be difficult to separate *H. habilis* from contemporary *Australopithecus*. *H. habilis* was succeeded about 1.6 million years ago by another type, *H. erectus* ("upright man"; the name is meaningless, since all known hominids move in an upright position), who is known to have lived also in Asia and possibly in Europe.

The main differences from *Australopithecus* appear in the construction of the cranium. The brain has increased, being on average close to 700 cubic centimeters in volume, while the face has reduced somewhat in size. An especially evident contrast is the relatively small shoulders, which do not protrude as in the robust *Australopithecus*. A comparison of the craniums of *H. habilis* and *A. boisei* provides a clear impression that we are contrasting a human with an animal. Although both are hominids, only *Homo* may be characterized as human, and the splitting of *Homo* from *Australopithecus* may be likened to the origin of humanity.

The physiques of the two beings do not evidence any big changes; naturally enough this is connected with the fact that even *Australopithecus* had a typically human way of moving.

Louis Leakey, Phillip Tobias, and John Napier were the first to describe *H. habilis,* on the basis of discoveries in the oldest layers at Olduvai (1.6 to 1.8 million years old). Well-preserved material has come from places east of

Lake Turkana (Koobi Fora); this species has also been unearthed at Omo and seems to have lived in South Africa as well. Infrequent finds at the Swartkrans cave show that *H. habilis* lived there at the same time as the robust *Australopithecus*. Whether *H. habilis* was prevalent outside Africa is so far uncertain. Finds from Java and southern China have with some doubt been ascribed to that species, but more recent investigations have indicated that they are younger. In Europe, on the other hand, shaped rocks and stones have been found that seem to date from the time of *H. habilis*. At Chilhac in France similar stones found in a stratigraphic layer below a hardened flow of lava date from 1.9 million years ago.

The oldest find thought to represent a predecessor of *H. erectus* in Africa dates from approximately 1.6 million years ago. This type is thought to have evolved from *H. habilis,* with which it differs, for example, in its larger brain size (on average, about one thousand cubic centimeters; actual size varies between approximately 800 cubic centimeters in the earlier types and 1,100 in the later) and in its smaller molars.

From these finds it becomes clear that *H. habilis* and *H. erectus* lived side by side with the later individuals of *Australopithecus* for a long time, probably more than a million years. According to an earlier theory (the so-called one-type theory), all hominids at each point in time have belonged to the same species, and evolution has thus proceeded on a "wide front" from the gracile and then the robust *Australopithecus* to *H. erectus*. This, however, cannot have been the case, since *H. erectus* has been found in some sediments that also contain *A. boisei,* and the difference between them is so marked that it is impossible to ascribe them to the same type. Thus, we can hardly regard them as members of the same species.

The branching of *Homo* from *Australopithecus* probably occurred when a local population of gracile *Australopithecus* changed its way of life and entered a new and different selection process, which in turn led it onto a new developmental track. It is of course impossible to identify with any degree of certainty what the difference was, but it probably was concerned with behavior, or what one would call "cultural" differences. This leads us to ask whether we can find any traces of cultural activities that diverge from what we already know about *Australopithecus*.

One answer is at hand. In sediments that accumulated between 2 and 2.4 million years ago we have found stone artifacts, i.e., rocks and stones shaped into rudimentary tools. This is also the time that we assume *Homo* evolved. Therefore, it is reasonable to suppose that this cultural innovation—working on stones to give them the desired shape, instead of being satisfied with what was available—became rather significant. In its simplest and earliest form the work consisted of one or more chips being hewn from a stone so that a sharp edge resulted. Such an edge might be used to slice open the skin of a dead animal to get to the meat. One is again reminded of Engels's theory that work is an important precondition for the attainment of the human state.

Still another change may have had a cultural import. We noted earlier that *Australopithecus* on the whole had a flat cranial base, indicating a lack of articulated speech. As mentioned above, in *H. erectus* an inward bend in the

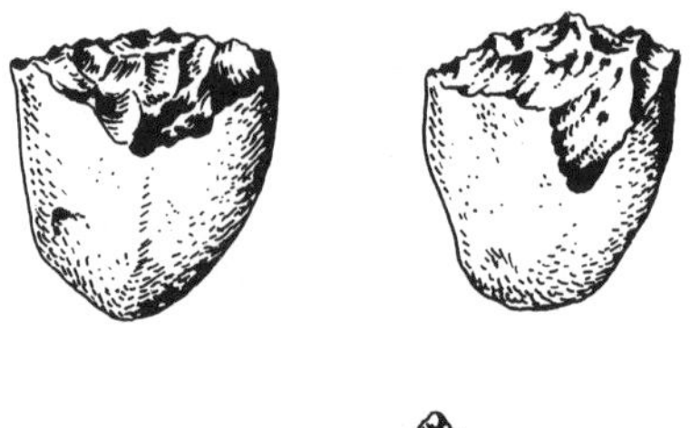

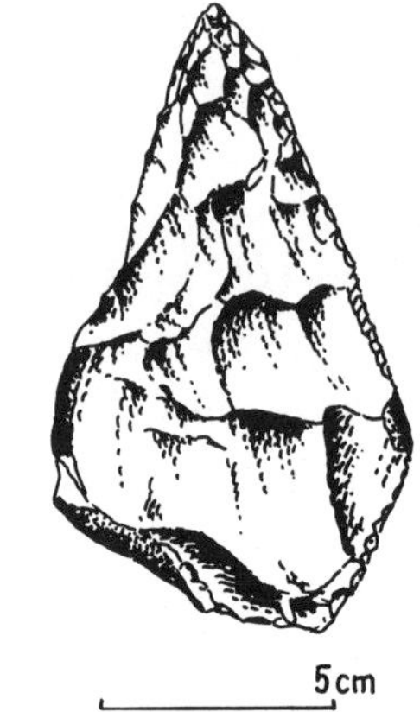

Figure 8.2 Stone tools. *Above:* roughly hewn cobbles, probably the work of *Homo habilis.* As the tools were used, chips presumably were broken off; the original tool most likely did not resemble the rock nuclei pictured here. *Below:* a primitive Abbeville axe. Tools made from cobbles appeared 2 to 2.5 million years ago, the Abbeville tools approximately 1.5 million years ago.

cranial base appeared approximately 1.5 million years ago, causing a deepening of the pharynx and, most probably, the ability to speak. This anatomical change took place long after the first *H. habilis* had appeared, and we should consequently keep in mind that the function of speech should have come before the anatomical adjustment. We can not expect that there would have been any pressure for selective change for a deepened pharynx before human beings had begun to speak. Therefore, *H. habilis* may have possessed a primitive kind of speech that was basically articulated with the aid of various consonants, more or less like a small child who commands only one vowel sound, eh, but nevertheless can say several words. This might spur the selection process to increase the power of expression through, for example, a deepening of the pharynx with an accompanying increase in the range of vowels. Such a cultural change may have been associated with the invention of the stone forge and probably with other innovations. Generally speaking, we may say that the development of humans from the time of early *Homo* has been dominated by cultural evolution.

When culture becomes a determinant of evolution, a great premium is placed on a growing intelligence. As we have seen, no real development occurred in the brain size of *Australopithecus;* the modest increase in the size of its braincase may be considered a consequence of increased body size from the gracile to the robust form. In *Homo,* on the other hand, a dramatic increase in braincase capacity clearly began with *H. habilis.*

Another anatomical change is the reduction in molar size. This may have been caused by a change in the diet. As we have noted, food composed mainly of vegetables and fruits seems to have played a major role in the diet of *Australopithecus.* On the other hand, the presence of stone tools with sharp edges may indicate an increased consumption of animal substances on the part of *Homo.* Even in this respect, it seems that *Homo* and *Australopithecus* went their separate ways.

A change that occurred within the genus *Homo* is the lengthening of late childhood and adolescence. This began at a late stage, and at present it is unknown whether it happened with *H. erectus* or later. A cultural connection is evident: this lengthening provides a long period of learning and preparation for adult life.

This change may be considered an addition to a developmental tendency that was in effect early among the higher primates and culminated in man. Called *neoteny* (or paedomorphosis), it implies that embryonic or "childish" characteristics tend to remain with a fully grown organism. The adult human being is in many respects childish, even in spiritual and mental characteristics. It has been said that a fully grown ape is a perfect bourgeois type, while the human at his best preserves his playful disposition and curiosity far into old age. Neoteny was without a doubt a significant developmental mechanism even when *Homo* branched from *Australopithecus*.

CHAPTER NINE

"The Ape-Man"

The creature called *Homo erectus* was discovered from 1891 to 1894 in Java by Eugene Dubois. He found the top of a primitive skull in addition to a thigh bone of a completely modern type and from this drew the conclusion that they had belonged to an upright-walking ape-man (*Pithecanthropus erectus*). Today the type is thought to be so closely related to modern man that it shouldn't be called a separate type and the name *Pithecanthropus* should be removed. However, the term *ape-man* has stuck despite the fact that it is misleading.

Not long ago it was established that the thigh bone of "Java man" is more recent than the top of the skull. However, other finds have confirmed Dubois's opinion

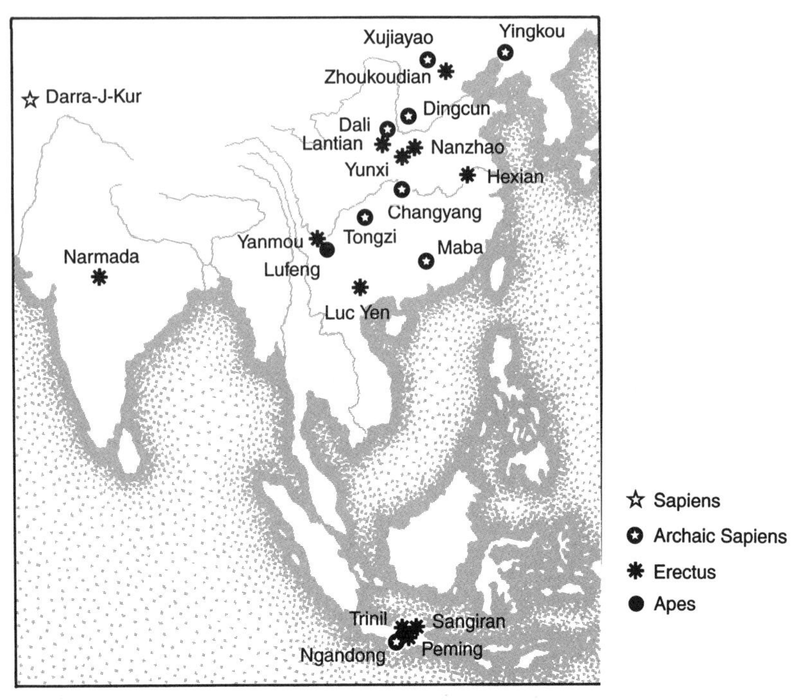

Figure 9.1 Fossil finds made in Asia of Ice Age primates.

that *H. erectus* walked upright. Since then, a rather comprehensive quantity of Java man remains has been unearthed at various localities in Java. In China the closely related "Peking man" was discovered in 1927 by Birger Bohlin, a member of a Chinese-Swedish expedition. The place where it was found, Zhoukoudian (Choukoutien), a cave, has yielded remains of more than fifty individuals; however, many disappeared or were lost during the Second World War. Remains of this human type have also been found in other parts of China—Lantian, Nanzhao, Yunxian, and Hexian in central China in addition to Yuanmou, perhaps the oldest of all, in southern China. Finds have also been made at Luc Yen in northern Vietnam and in the Narmada Valley in central India.

Homo erectus has also lived in North Africa (Ternifine in Algeria, Sidi Abderrahman in Morocco), in the area around Lake Turkana (for example, Koobi Fora), at Olduvai, and at Olorgesailie in Tanzania. This primitive human has also been found in the South African grottoes Sterkfontein and Swartkrans, but these remains seem to appear in younger sediments than those containing *Australopithecus.*

Even European finds have been classified as *H. erectus;* we will discuss them in the next chapter.

The ages of these finds vary between 1.6 and 0.2 to 0.3 million years; Peking man from Zhoukoudian belongs to the youngest groups. This grotto is thought to have been inhabited between 460,000 and 230,000 years ago. Earlier determinations claimed that certain South Chinese and Javanese finds are as much as 1.8 to 1.9 million years old. Revised datings indicate that they are considerably younger, not more than 1.2 million years old. I should point out here that the oldest finds made in Java seem different because of their unusual size. They were originally described under the rubric *Meganthropus* by their discoverer, G. H. R. von Koenigswald. Certain scholars have classified them as *Australopithecus;* others have placed them in the same group as the relatively big-toothed *H. habilis.* Their status is still somewhat unclear.

During the long period in which *H. erectus* lived, a noticeable development took place. The average volume of the cranium among the early Javanese specimens was

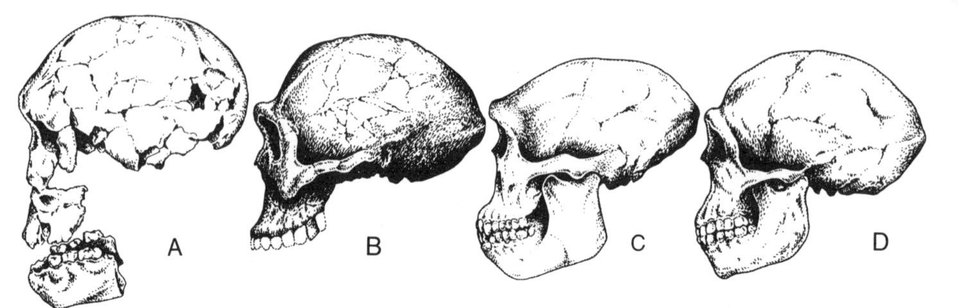

Figure 9.2 Skulls of early *Homo*. (*A*) *Homo habilis,* East Africa, about 2 million years old. (*B–D*): *Homo erectus* (*B*) from East Africa, about 1.5 million years old; (*C*) from Java, about 1 million years old; and (*D*) from China, about 0.4 million years old.

800 cubic centimeters; among the later specimens unearthed in China it was 1,060 cubic centimeters; in an equally late population from the Solo River in Java it was 1,090 cubic centimeters. An extensive cultural development occurred at the same time. *Homo habilis* had made simple tools from boulders. With *H. erectus* the tools became much more effective and varied according to intended use. *H. erectus* of the later period had already made use of fire, as, for instance, at Zhoukoudian.

The face shows a characteristic feature: the especially well-developed protuberance above the eyes, which runs from one side of the face to the other without any interruption above the nose. A similar protuberance is also present in *H. habilis* and in *Australopithecus,* but it is noticeably more substantial in *H. erectus,* which even has heavier bone in its cranium, especially among the Asian types. Among these types, the protuberance is practically straight, while among the African types it tends to bend upward above the eye sockets, a feature that was also present among the people of the European Ice Age. Thus there were some dissimilarities between the Asian and the African *H. erectus.* All specialists now tend to consider them two closely related but separate types.

In many individuals of *H. erectus,* especially the early ones, the forehead is almost nonexistent. This is especially true of the early Asian finds. In the later Asian types, as in the African ones, the forehead rises more precipitously. However, the cranium is low and long and, viewed in profile, extends to a point in the back. The braincase is widest nearest its base, the opposite of *H. sapiens,* in whom it is widest in its upper part.

The face has prominent jaws. The molars are smaller than those of *H. habilis,* but still much more powerful than ours. The chin is indistinct and has no protuberance demarcating it.

The rest of the skeleton is incompletely known. On the basis of the bones that are known, scientists have concluded that body height was rather modest, about 160 centimeters. An almost complete skeleton discovered in 1984 by Kamoya Kimeu west of Lake Turkana is consequently quite significant. So far, however, we have only a preliminary description. The remains are of a twelve-year-old boy (using the age criteria for *H. sapiens*) whose height was approximately 160 centimeters; it has been calculated that he would have been about 180 centimeters tall as an adult. This is an impressive height for *H. erectus* and may be a trait which further differentiates the African from the Asian type, although it is not surprising to find local types that diverge from one another in body height. The same is true of humans of today. The bones in the skeleton are, at least among the later *H. erectus*, very much like those of contemporary humans and indicate a rather slender physique.

The cultural sequences, as manifested in the stone tools, are especially well documented in East Africa. The oldest one, known as the Oldowan, is the "boulder culture" most closely associated with *H. habilis*. It was succeeded by two cultures typical of *H. erectus*, namely, Abbeville and Acheul. As the names indicate, these cultures were originally characteristic of France. What distinguishes them most are the hand axes of the Acheulian model, which are made of flint or are cut from flintlike rocks with a glassy consistency, as, for instance, lava. The flint occurs as irregular nodules or layers in limestone, and it consists mainly of silicic acid. By cutting away pieces, one can produce an edge just as sharp as a glass edge. What these *H. erectus* tried to make was a pear-shaped hand axe with one pointed end and with sharp edges along both sides. At the Abbeville stage, the oldest one, the pieces have been cut away roughly, the edges are uneven, and part of the rock's original surface may still remain. The Acheul axe is an improved version:

Figure 9.3 Acheul hand axe.

the pieces have been cut away more accurately, edges are more even and the surface is completely worked over. In Africa, lava is the typical raw material.

The actual use of these hand axes has been widely debated. They may have been a universal tool used in cutting, drilling, and chopping. In an interesting experiment an Acheul axe has been thrown much like a discus. It flew for a distance but then fell almost vertically with the sharp point facing downward. Speculation has therefore been made that it was used as a hunting weapon more or less like a boomerang. Another view has it that the base of the axe was mounted to a "handle" of mastic (a mixture of resin and clay). It has also been surmised that the axe was lashed to a wooden handle.

In time an increasing selection of specialized tools was added to the hand axes. All stone tools have a short life, since their cutting edges are worn down rather quickly. Repeated sharpening may restore an edge, but the entire shape of the tool will eventually diverge from the original one, and become useless. It is no wonder that, at such sites as Olorgesailie, we find hundreds of thousands of hand axes.

At Olduvai, the Oldowan culture was succeeded by the Abbeville without any transitional phase. Abbeville was a long-lasting culture, evolving into the Acheul 0.5 to 0.8 million years ago.

The Asian cultures are less well known. The earliest are of the "hacked-stone" type. The late Zhoukoudian culture shows very careful work but is atypical, since flint was not available and everyone had to be satisfied with quartzite.

Late African *Homo erectus* has often been considered an "archaic *Homo sapiens*" or a transitional form between *Homo erectus* and *Homo sapiens.* "Rhodesia man" from Kabwe in Zambia belongs to this group, as do similar finds from Elandsfontein at Saldanha Bay and the Härd grotto at Makapan (both in South Africa), as well as finds from Lake Eyasi in Tanzania. Their ages are in dispute, but they are thought to be more than 100,000 years old, perhaps 200,000 or more. Associated cultures were evidently later than the typical Acheul. Braincase volume is definitely larger than that of a typical *H. erectus,* exceeding 1,250 cubic centimeters. Despite these advanced charac-

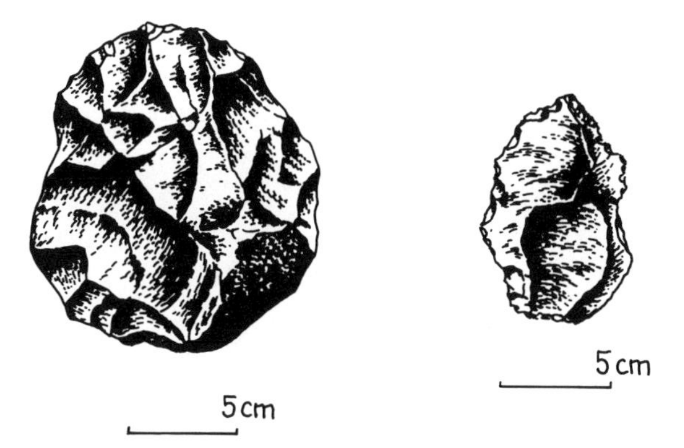

Figure 9.4 *Left,* cut stone from the culture of Java man. *Right,* stone point from the grotto of Peking man.

teristics, the head has a marked likeness to that of *H. erectus,* with thick eyebrows that run together at the base of the nose, a large face with projecting jaws, and a braincase having its maximum width at its base. The skeleton is otherwise like that of *Homo sapiens,* and Rhodesia man thus diverges from the European Neanderthal, with whom it has also been compared.

The Kabwe find is distinctive because the teeth have been seriously damaged by caries, which otherwise are unusual among fossil humans. The most ordinary physical signs of damage suffered by *H. erectus* are marks made by blows, especially among the Asian types with their thick cranial bones. A peculiar disease has been noted in an *H. erectus* from Koobi Fora, namely, vitamin A poisoning. Vitamin A collects in the liver and may attain a concentration that is poisonous to human beings. Did the unlucky human from Koobi Fora have lion liver as his favorite meal?

It is now universally agreed that our own kind, *Homo sapiens,* has evolved from *Homo erectus*. The later African types with their large brains and their modern-looking skeletons undeniably appear to be the ancestors of modern human beings.

CHAPTER TEN

Humans in the European Ice Age

The Ice Age in Europe began about 1.6 million years ago. It was not thoroughly frigid during the entire period. Icy "glacial" periods, during which a huge inland glacier covered northern Europe, with smaller ice fields covering the Alps, the Pyrenees, and other upland areas, alternated with warmer "interglacials," during which the climate was as warm, or warmer, as it is now. The interglacial periods, relatively brief episodes (10,000 to 15,000 years), returned at intervals of approximately 100,000 years. "Interstadials" occurred within the interglacial periods, during which the climate became somewhat milder, although not as warm as during an interglacial period. The last interglacial period (the Eem interglacial) started about

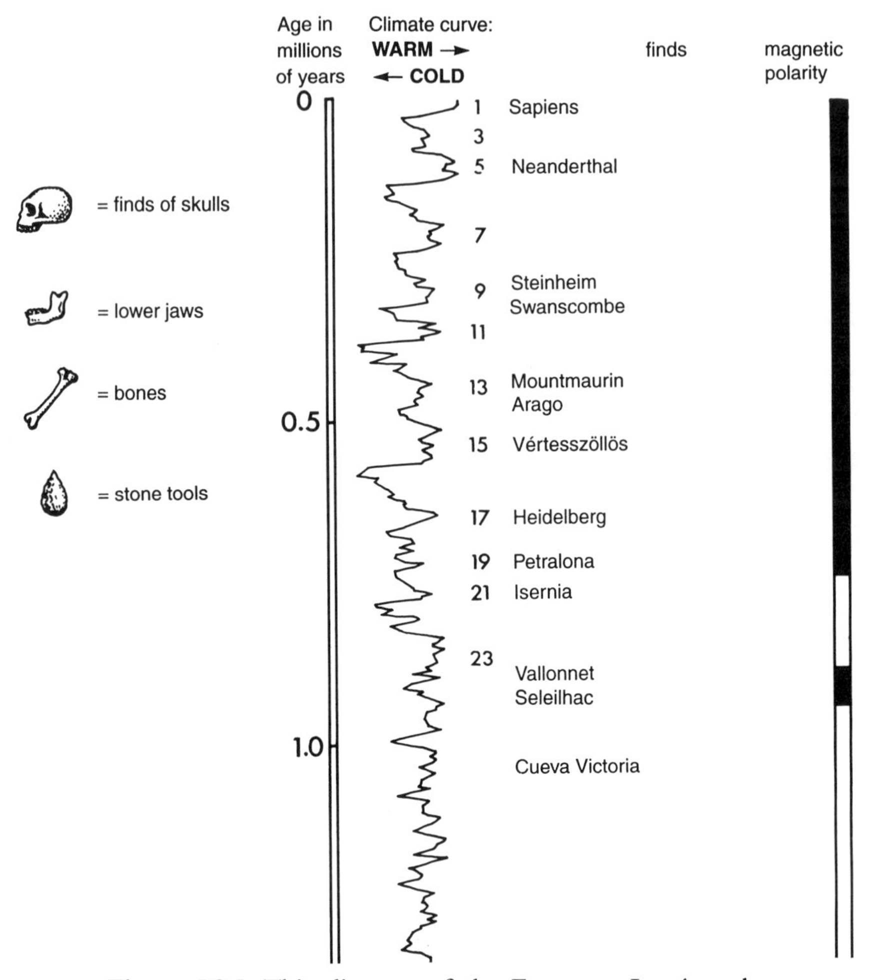

Figure 10.1 This diagram of the European Ice Age shows changes in climate and magnetic polarizing, important finds of fossil humans, and typical cultural remains. Earth magnetism is normal (black) or reversed (white; the North Pole becomes the South Pole, and vice versa). During the last million years one can differentiate between twenty-four climatic stages, warm (odd numerals, 1–23) or cold (even numerals, 2–24). Only the warm stages are indicated in the diagram.

120,000 years ago. During an interstadial period 30,000 to 40,000 years ago, a major part of the inland ice melted, and most of Scandinavia became ice-free. The interglacial period in which we are now living began approximately 10,000 years ago.

The Ice Age in Europe coincided with the Pleistocene era in the geological chronology. It is generally agreed that the Pleistocene ended and the present epoch (Holocene) began ten thousand years ago. Actually, the Ice Age is still going on, and a new interglacial period may occur in the future.

The oldest traces of humans in Europe seem to be stone artifacts from the late Pliocene period, as, for instance, those found at Chilhac, which are thought to be 1.9 million years old. Except in southern Europe, there are no signs of human beings on the European continent from the early Pleistocene period. Finds in Spain (the Victoria grotto, about 1.3 million years old) and in Yugoslavia indicate human habitation. However, it was not until 0.9 to 1 million years ago that human beings once again immigrated into Europe. The worsening climate may have driven away the former immigrants, or they may merely have died out. The immigration one million years ago, on the contrary, seems to have led to permanent settlement, and one can assume that by then humans had developed a culture that made it possible to exist in this frigid climate.

No skeletons of these early Europeans have been unearthed, but artifacts in sediments that can be handily dated—as, for example, at Soleilhac and Vallonnet in France—testify to their presence. A handful of finds from Petralona in Greece, Heidelberg in Germany, Vértesszöllös in Hungary, and Arago and Montmaurin in the French Pyrenees date from somewhat later, probably between 400,000 and 800,000 years ago; they have been listed here in the most likely chronological order, from the oldest to the youngest. The Petralona find consists of a well-preserved cranium; from Arago we have the remains of several individuals, including parts of a cranium. As a precaution they are usually designated "archaic *H. sapiens,*" a noncommittal name if there ever was one. One characteristic is an especially robust physique. The skeletal bones that have been found are enormously powerful, and the bone of the skulls is as thick and heavy as that of the contemporary *H. erectus,* ranging from 1,250 to 1,500 cubic centimeters. The eyes are not shaded by a continuous thick brow as in the African *H. erectus;* in this case it is severed in the middle by the nose ridge. In all these respects similarities exist with later Europeans (the Neanderthals), but many characteristics also remind us of *H. erectus,* and certain scholars prefer to assign them to that family.

A somewhat younger group of finds, including those from Atapuerca and Cova Negra in Spain, Lazaret in southern France, Steinheim and Bilzingsleben in Germany, and Swanscombe in England, are probably between 200,000 and 400,000 years old. These finds tend toward a more gracile physique with brain volumes between 1,200 and 1,300 cubic centimeters. Similarities have been noted with both Neanderthals and *H. sapiens,* predominantly with the former. Again, the large brain volume is a sharp contrast to the contemporary *H. erectus.*

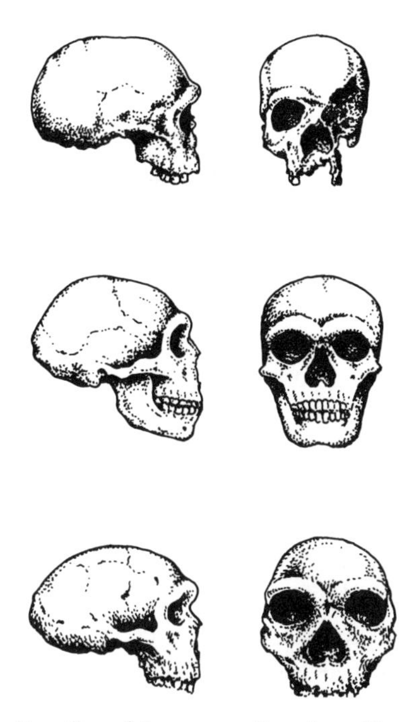

Figure 10.2 Fossils of humans: Ice Age Petralona, Greece, about 700,000 years old; Steinheim, Germany, about 300,000 years old; Neanderthal man, about 40,000 years old.

Even this group of finds is usually designated archaic *H. sapiens*.

Generally, the older group is associated with a culture of the Abbevillian type, while the younger one represents the Acheul culture. Aside from these axe cultures, so-called slab-axe cultures (clacton and others) also appear, in which the hand axes were not made from the "core" of the flint nodules but were cut from a big slab.

From the beginning of the Eemian interglacial period

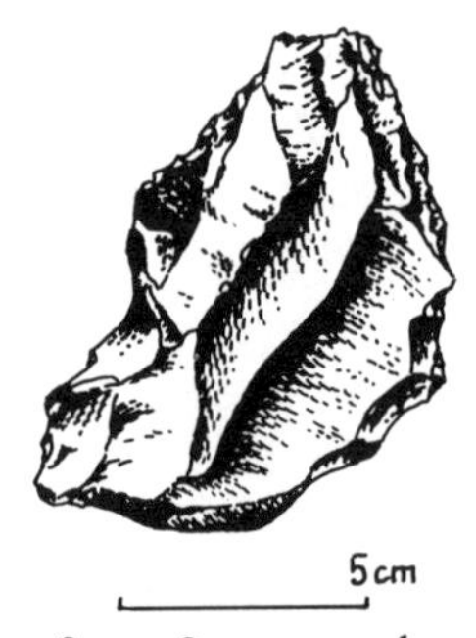

Figure 10.3 Slab axe from Swanscombe.

120,000 years ago to approximately thirty thousand years ago, Europe was inhabited by the Neanderthal people, the best known of all the fossil human types. We know so much about them mostly because the Neanderthals buried their dead, and whenever this took place in grottoes, the skeletons have often been preserved until today.

The discoverer of Neanderthal man and the founder of paleoanthropology was the schoolteacher Carl Fuhlrott from Elberfeld. In 1856 he made public his discovery at Neanderthal near Düsseldorf, maintaining that it was an extinct human type. (Two older finds, at Engis in Belgium and at Gibraltar, were known, but their significance had not been realized.)

While the finds from the Eem interglacial period are relatively few (for instance, at Saccopastore in Italy, in a cave in the Rock of Gibraltar, at Taubach and Ehringsdorf near Weimar in Germany, and a large number of individuals at Krapina in northern Yugoslavia), Neanderthal people from the latest glacial period have been discovered in almost one hundred places, their ages varying from one hundred thousand to thirty-two thousand years. Many discoveries have been made in France, but Neanderthal remains have also been unearthed in Portugal, Spain, Switzerland, Germany, Belgium, Czechoslovakia, Hungary, Romania, Italy, Russia, Iraq, Lebanon, Jordan, Israel, Libya, and Morocco. Even though Neanderthal man was a European phenomenon, he has been found in Western Asia and in North Africa.

Most finds have been in caves, which in many cases served as habitations. However, pieces of tent poles show that Neanderthals put up some kind of tent inside the caves (this was done by even earlier European peoples). The relative abundance of finds in caves is not because Neanderthals were "cave dwellers," but because finds made in caves stood a much better chance of being preserved than remains left under the open sky.

The Neanderthal was a unique type of human being. The head was large and long, and the braincase (despite the thick bone in the cranium) was enormous, with a total capacity of 1,270 to 1,750 cubic centimeters. In the male, the average head size was about 1,600 cubic centimeters; in the female somewhat over 1,300. The forehead was overshadowed by the ridges over the powerful eyebrows. The eye sockets were large and round. The bridge of the nose was pronounced, the nostrils big and round, and the jaw protruding. The cheeks were not sunken (with pronounced cheekbones) as in modern humans. The teeth were large, and the roots of the molars were partly grown together. The chin hardly protruded but occasionally a trace of a protruding chin has been found, as is the case with us.

The physique is especially noticeable because of its great

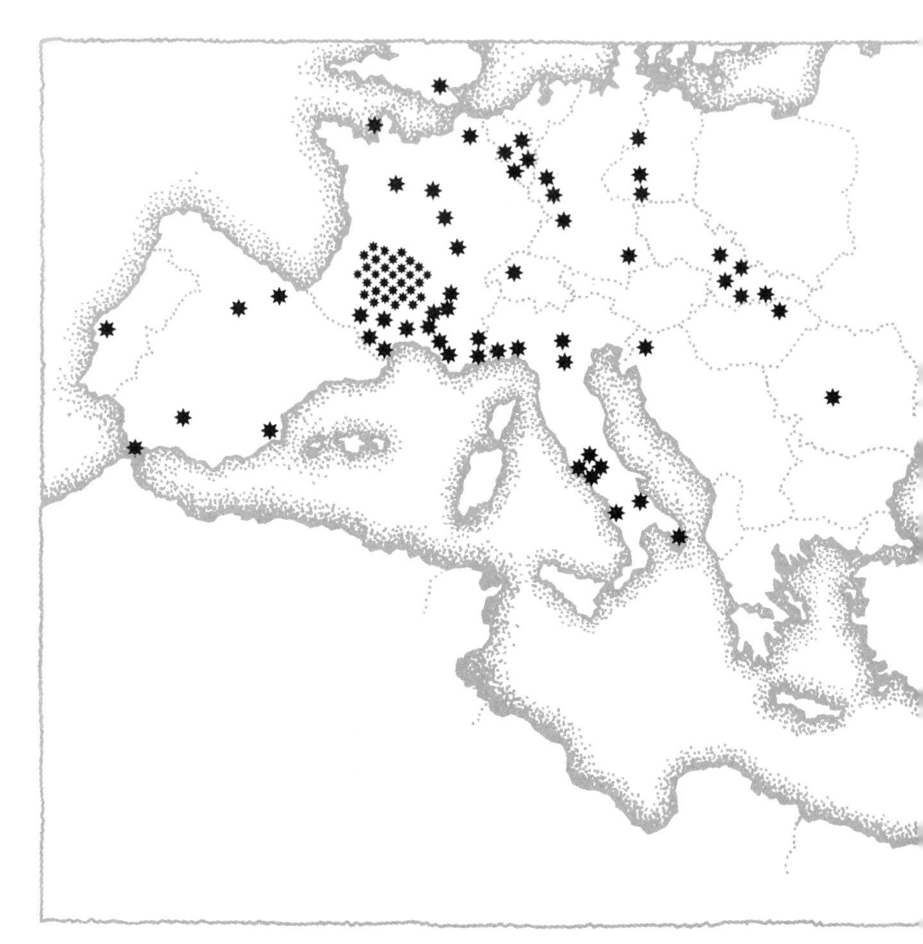

Figure 10.4 Localities where remains of Neanderthals and their precursors have been found.

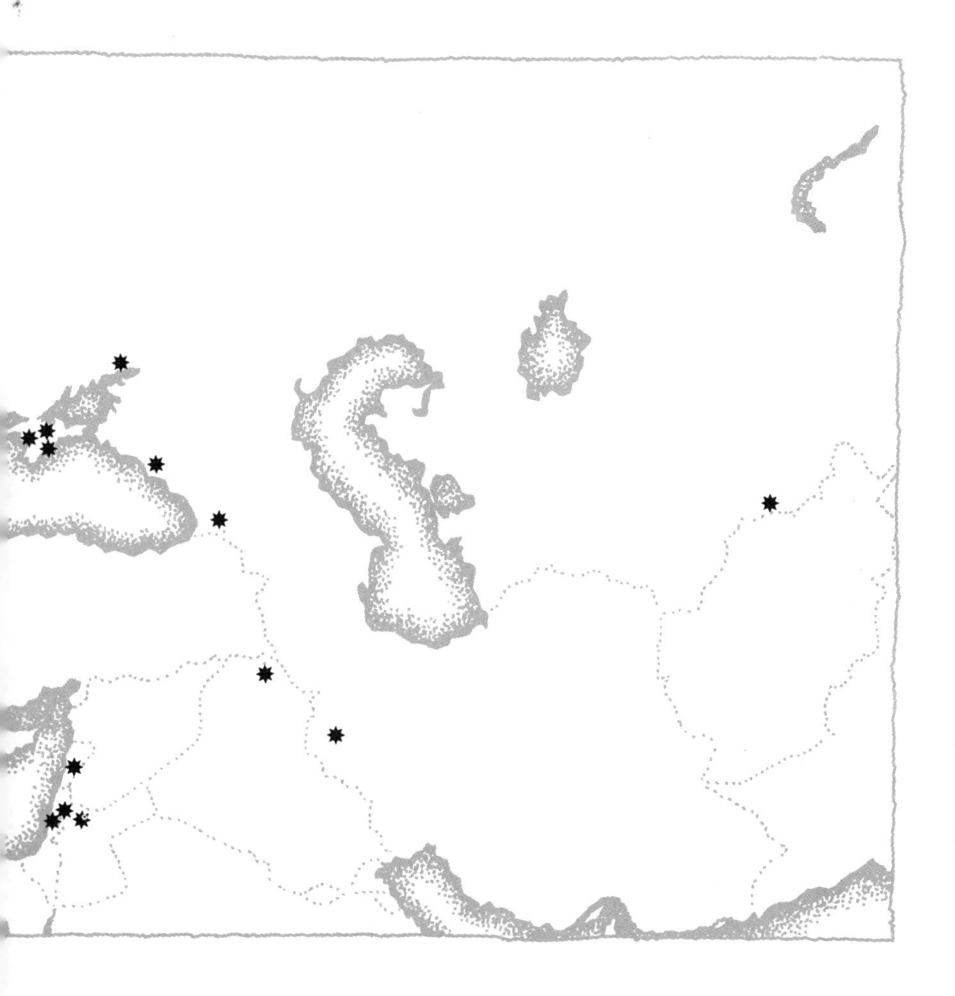

robustness; the shoulders were very wide. The lower parts of the arms and legs were comparatively short (as with Lapps and other peoples adjusted to a frigid climate). The two long bones of the lower arm, the radius and the ulna, were further apart than is the case with humans. The hands were large. The hips were wide, even in the male. The thigh bones were big and thick and a bit curved. Even though the average height of the body was only about 168 centimeters in the male and 159 centimeters in the female, the body build indicates a large bulk and an average male weight of perhaps one hundred kilos. The size of the brain may thus have been a consequence of the body's great weight.

The unusual hip width may at first seem surprising; this causes a more pronounced knock-kneed stance, which mechanically is disadvantageous. On the other hand, wider hips served to increase the width of the birth canal in the female, and an authority on Neanderthals, Erik Trinkaus, regards this as the reason. The fact that even the males had broad hips speaks to an evolutionary byproduct. Trinkaus believes that pregnancies were probably longer than today, perhaps eleven to twelve months, and that the child consequently was advanced at birth. The same may have been true of *H. erectus,* who, however, had a smaller head and thus was in no need of a wider birth canal. The problem of large heads and narrow birth canals has been solved by the child being born at an earlier stage.

The earlier (interglacial) Neanderthals did not evidence features as extreme as the later, or "classical," Neanderthals of the latest glacial period. Among other things, the braincase was somewhat smaller, ranging from 1,200 to 1,450 cubic centimeters among four individuals; it should be noted, however, that three were females, thus the difference is not very big. Even in these creatures broad hips are typical.

The typical culture of the Neanderthal people is the Mousterian, a slab-axe culture marked by a large selection of various tools, among them stone lance or spear points. Wooden spears were also used, their points hardened by fire. Neanderthal man used fire for other purposes also; animal bones believed to be left over from meals often show signs of having been put in the fire, and fossil hearths are found in many caves. Neanderthal man conducted rites that may be described as religious: the dead were buried with funeral gifts (bones with meat on them, flowers, in one case butterflies). Examples of cannibalism do exist, especially from the Eem interglacial period (Krapina); later, the brain was eaten, something that also oc-

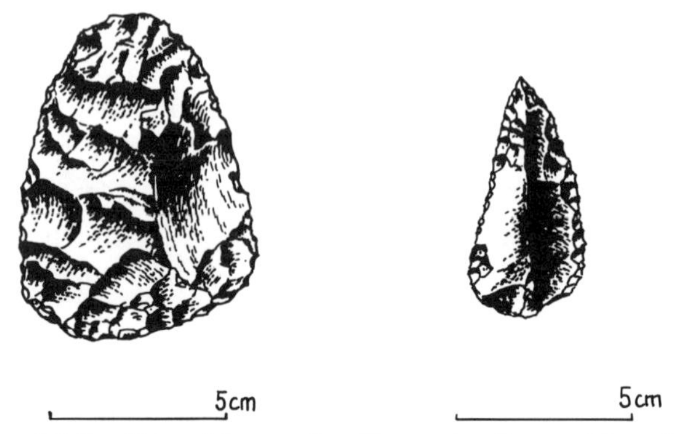

Figure 10.5 *Above:* Mousterian flint point from a Neanderthal find. *Below:* Mousterian slab axe.

curred among *H. erectus* (Zhoukoudian, Solo) and European pre-Neanderthals (Steinheim). This possibly had a ritual significance.

The Neanderthals enjoyed long lives. In an extensive find in Hortus Cave in southern France more than 20 percent of the people had been over fifty years old. Many seriously handicapped individuals lived long and were evidently taken care of by their tribe (for example, a toothless and rheumatic man at La Chapelle-aux-Saints in France; a man with a withered arm at Shanidar in Iraq).

Neanderthal man had many characteristics that we regard as thoroughly human. In most classifications he is correctly described as a subspecies of us: *Homo sapiens neanderthalensis*. There is a current trend in paleoanthropology toward widening the definition of our species so that it includes older human types, even *H. erectus*. William Howells has this to say in this regard: "To doubt this looks as if one accepts a kind of evolutionary racism. . . .

A writer even maintains that all kinds of anthropology are racist and anti-humanistic as soon as we classify, differentiate, or objectify. Before we agree to abandon the study of human variation and rush into pre-Linnaean innocence, trampling on poor Mendel and Darwin, we must ask ourselves precisely how does one set about studying evolution. . . . Through humanistic revelation?" Actually, the anatomical differences between Neanderthal man and us are as great as those between two types of mammals, such as the cave bear of the Ice Age and its cousin, the brown bear, which also comes down to us from the Ice Age but is still with us.

Neanderthal man disappeared about thirty thousand years ago. One of the youngest fossils, from St. Césaire in France (thirty-two thousand years old), is, interestingly enough, associated with an advanced culture of the "lower paleolithic type," to which we shall return in the next chapter. By then modern human beings already lived in Europe, and some scholars believe that they lived side by side with the Neanderthals, with a rather stable boundary between them, over several thousand years. The earlier view holding that our kind evolved directly from Neanderthal man must now be regarded as extremely improbable. There is no reason to maintain that present-day humans originated in Europe, and Neanderthal man is so often made to appear as our direct ancestor solely because of a historical accident—he was discovered early, and known fossil material is quite extensive.

CHAPTER ELEVEN

Us

In the professional literature we often find the term *the anatomically modern* Homo sapiens. It refers to people of the present day in contradistinction to the so-called archaic *H. sapiens.* For the sake of simplicity we will use the designation *H. sapiens* to refer only to this "anatomically modern" type.

H. sapiens differs from all other human types by having a high, rounded braincase and also by lacking a faintly shaped or bony curvature above the eyebrows; the face is vertical without protruding jaws; the chin is well developed; the teeth are smaller. Furthermore, *H. sapiens* has a larger brain than *H. erectus* and a more slender physique and a differently shaped pelvis than Neanderthal.

The oldest finds of *H. sapiens* occurred in Africa. In strata about 130,000 years old at Omo in Ethiopia, Richard Leakey found the remains of three individuals with predominant *H. sapiens* traits but with other characteristics similar to those in *H. erectus*. These craniums can be considered *H. sapiens* even though the bony curvature above the eyebrows is more marked than in modern humans. This was the beginning of a series of African finds between 100,000 and 35,000 years old, all of *H. sapiens*. Many finds were made in South Africa, in such places as Klasies River, Mouth Cave, Border Cave, and Florisbad; found at Florisbad was a cranium approximately 100,000 years old, which despite relatively large eyebrow bones indubitably is *H. sapiens*. Even though these age determinations are not quite definitive, it seems clear that *H. sapiens* lived in Africa much earlier than in Europe.

Even in other areas the modern *H. sapiens* seems to have appeared relatively early. The cranium of a teenage *H. sapiens* discovered in Niah Cave in Sarawak (Borneo) has been dated at 40,000 years old. It should be noted that the large Sunda Islands had close connections with Asia during the Ice Ages, when huge quantities of water were bound up in the inland ice and ocean depth was more than 100 meters lower than today. Early humans invaded Australia at about the same time. These primitive people must have been able seagoers, since the transoceanic voyage, in spite of the lower water levels, was long. Maximal glaciation 18,000 years ago made the distance (across the Timor Sea) a mere ninety kilometers, but 40,000 years ago it was probably twice as long.

Exact dates for the appearance of the earliest *H. sapiens*

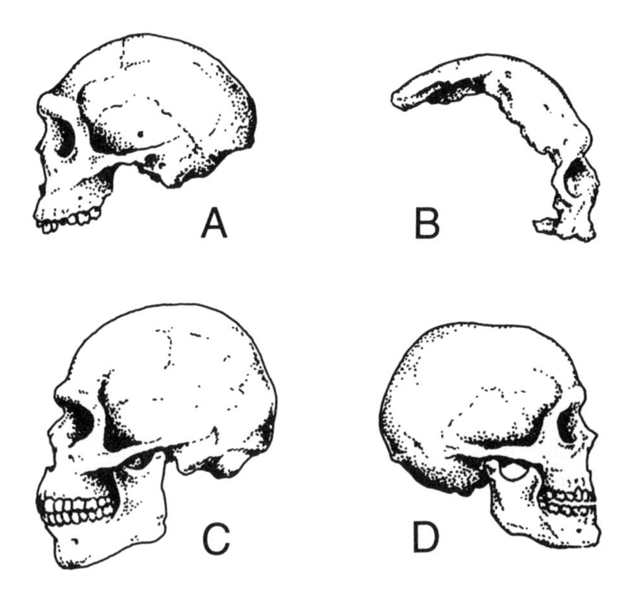

Figure 11.1 A probable sequence of development from *Homo erectus* to *Homo sapiens* begins with (*A*) the so-called Rhodesia man from Kabwe (age 400,000 to 200,000 years). At (*B*), a find from Florisbad in South Africa (age 200,000 to 100,000 years), we approach the anatomically modern *Homo sapiens.* (*C*) is a typical *Homo sapiens,* from a grotto at Mount Carmel, Israel (age about 35,000 years). (*D*) is an Ice Age European Cro-Magnon man (age 30,000 to 20,000 years), a typical member of the "white" race.

in East Asia are unknown. In Europe, the first members of *H. sapiens* appeared about 35,000 years ago, but the oldest definitively dated find is somewhat younger (at Kelsterbach in Germany; 31,000 years old). In 1868 five Ice Age skeletons of *H. sapiens* were found at Cro-Magnon in Les Eyzies, and all of Europe's Ice Age *H. sapiens*

Figure 11.2 Flint chip with parallel sides, worked at both ends. From a Magdalenian find.

have been named for that place. Actually, Cro-Magnon man is much like a contemporary European, on average having a somewhat longer cranium and being a bit taller. The differences, however, exist merely in the average measurements and are completely overshadowed by individual variations.

H. sapiens is associated with the lower Paleolithic cultures; the most important ones in Europe, listed chronologically from oldest to youngest, are as follows: Aurignacian, Gravettian, Solutrean, and Magdalenian. This sequence took place between 35,000 and 10,000 B.C. and is characterized by an ever-increasing wealth of tools, hunting weapons, jewelry, and other objects. Typical of lower Paleolithic cultures was the making of long, narrow flint chips with parallel sides and a sharp point at one end.

One of the earliest lower Paleolithic cultures, a contem-

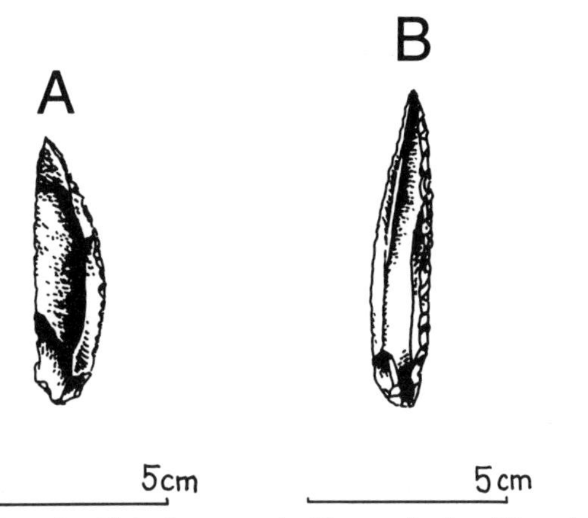

Figure 11.3 (*A*) Châtelperron knife made by Neanderthals from a flint chip; (*B*) Gravettian knife made by *Homo sapiens*.

porary of the Aurignacian, was the Châtelperron, a Neanderthal culture. Although it produced flint chips (for instance, the beautiful Châtelperron knife), it also seems to derive from the Mousterian culture, an early Neanderthal culture. This shows, however, that the Neanderthals did not lack for innovations. Part of the Châtelperron heritage also seems to be present in the Gravettian culture (among them the flint-chip knife).

The Aurignacian culture, which with Châtelperron launched the lower Paleolithic period, introduced carving and shaping of bone objects—for example, light spear tips. It was followed by the Gravettian, which was prevalent even in Eastern Europe, where mammoth hunting was a specialty.

Succeeding the Gravettian period in Western Europe was the Solutrean, famous for its beautifully formed flint

Figure 11.4 "Microliths" from the end of the Ice Age. They were mounted as barbs or saw teeth in wood or bone tools.

spear tips; they were made with new methods, in which chips were pressured rather than hammered loose from the flint. These so-called laurel leaf tips were too thin to be used for any but ritual purposes.

The West European sequence of developments concluded with the Magdalenian culture, which shows some similarities to Eskimo culture. While the working and shaping of flint objects had a practical purpose, bone and horn objects often turned out to be exquisite works of art.

Carvings on cave walls and on objects in general originated in the Aurignacian period; an ever-richer tradition of art continued into Magdalenian times and seems to have lasted until the end of the Ice Age. Lower Paleolithic *H. sapiens* embellished the famous painted caves, at, for instance, Lascaux, Font de Gaume, and Altamira. Although most of the paintings depict animals, there are also pictures of human beings and plants, in addition to abstract signs and series of different signs. In general,

Figure 11.5 One of the famous exceedingly thin "laurel leaf points" from Solutré in France, length about nineteen centimeters.

lower Paleolithic culture was highly developed, artistic, and surprisingly rich in a purely material way.

Neanderthal man was succeeded by *H. sapiens* even in the Middle East, but in the beginning no cultural change took place here. The early representatives of *H. sapiens* carried forth a so-called Levallois-Mousterian culture similar to that of the Neanderthals who had preceded them. The oldest *H. sapiens* in this area—for example, those who

Figure 11.6 Magdalenian harpoon made of reindeer bone.

lived in the caves of Mount Carmel—were of the Cro-Magnon type, but in certain cases had somewhat bigger eyebrow ridges than the typical European representatives (although even in Europe similar examples have been found). The possibility that crossbreeding occurred between the Neanderthals and *H. sapiens* at Carmel has been proposed, but this is contradicted by the skeleton being in every other way a typical *H. sapiens* without any trace of Neanderthal characteristics. On the whole, no traces of Neanderthals can be found in Cro-Magnon man's anatomy in Europe.

The origin of *H. sapiens* has not been determined conclusively. For now it seems as if the origin must be in

Africa, where conceivable precursors are known (the advanced late *H. erectus* of the Rhodesia type); in all seriousness, there is hardly any reason to put forth any other alternative. But it should be kept in mind that much less is known of the Asian sequence of developments than of the African and that relatively advanced *H. erectus* types are known even there. In Europe, too, there were primitive types that might be considered, not Neanderthals but predecessors of Neanderthals—the older group (at Petralona and Arago) as well as the younger (at Steinheim and Swanscombe). Both these alternatives are conceivable, although less probable.

An entirely different theory about the origin of *H. sapiens* is the so-called polycentric theory. It holds that *H. sapiens* evolved among local populations, each of which gave rise to a separate race—Neanderthals to the Caucasoid ("white") race in Europe, Peking man to the Mongol race in East Asia, Rhodesia man to the blacks in Africa, etc. In its extreme form the theory is biologically improbable, and for Europe it is exaggerated, as shown by finds of *H. sapiens* outside Europe that are ascribed to a period earlier than that of the Neanderthals. But immigrating members of *H. sapiens* may have assimilated gene material from earlier primitive inhabitants in various quarters (even if this cannot be proved in Europe). For instance, it is possible to show some "Mongolian" features in the construction and shape of the teeth of Peking man, which might indicate a certain continuity.

Although there is a variety of races among the present-day population, they all belong to the same species: *Homo sapiens*. Well-known writer and scholar Stephen J. Gould has phrased it like this: "All of us who now live on the earth are brothers." We can identify three main races—the Mongoloid, the Negroid, and the Caucasoid—in addition to a number of smaller ones (that is, races with only a few representatives still living), such as the Australoid. Although the races differ from one another anatomically as far as averages are concerned, all differences are bridged through individual variation. No one has been able to prove that any differences exist among the races in mental characteristics. In many cases physical dissimilarities could have been adaptive and may have developed during the immigration stage—Cro-Magnon is a typical Caucasoid; the early *H. sapiens* in China were Mongoloid, etc.

An evident adaptive difference is skin color. Strongly colored skin pigmentation provides protection against the sun's ultraviolet rays, which if received in powerful doses may cause skin cancer; a dark complexion is therefore advantageous in the tropics. In the higher latitudes no such danger exists; instead, there is the risk of suffering from a lack of D vitamins, which we synthesize under the influence of the ultraviolet rays. A lack or shortage of D vitamins results in rickets or in osteoporosis (brittle bone construction). In Europe, therefore, the lack of pigmentation is an advantage. As a consequence of this, we can observe a gradual increase in pigmentation in Caucasoid groups ranging from Northern Europe to Arabia, India, and Sudan.

The last continent to be settled by humans seems to be America. It is usually assumed that humans immigrated

into North America via the Bering Straits, which were dry land during the ice ages. After their arrival in Alaska from Siberia, the immigrants encountered the inland ice, which extended as a barrier across the entire continent from the Pacific Ocean to the Atlantic. Approximately fifteen thousand years ago the ice had melted enough so that a "corridor" had opened to the south. The humans who lived in North America from 13,000 to 10,000 B.C. are known as Paleoindians; they were the bearers of an advanced big-game-hunter culture. They seem to have lost no time pressing southward and about eleven thousand years ago had already reached the southernmost parts of South America.

It has also been posited that North America was first colonized by people arriving from the sea, as was the case with Australia, but there is no proof for this assumption. On the other hand, some scholars believe in a much older human habitation in America, as much as 100,000 years or more. The problems, however, are debated, and the question is far from settled.

The most distant outposts of South America having been reached, humanity had completed its conquest of the inhabitable earth. The human beings who lived at that time, at the end of the Ice Age about ten thousand years ago, were in all essentials identical to us.

CHAPTER TWELVE

The Present and the Future

Predictions are most easily made by observing what has happened until now and extrapolating the trends forward in time. Humans can look forward to a further decrease in dentition (for example, loss of the wisdom teeth); an increase in brain size; and perhaps the loss of the little toe. But such a prediction is too simple; we have no guarantee that evolution will continue in the same direction it has up to now. When evolution reaches a certain "plateau," it comes to a standstill, as many examples show. At one time in our prehistory, humans changed from a four-legged to a two-legged stance—in this case the plateau was reached four million years ago, and what has happened since has been at most minor adjustments. Evolution may also change

FUTURE FOSSILS

direction. In *H. erectus* the bone in the skull grew thicker and heavier; in *H. sapiens* it became thinner again.

Are humans subject to change even now? The answer is yes. I don't refer to the increasing height of men and women in Western countries and in Japan, which is mainly due to improved nutrition; this is not evolution in the genetic meaning of the word. Evolution takes place when the population of certain groups of people (for instance, the Europeans) stagnates or decreases while that of others (in the developing countries) increases. To the extent that these groups differ genetically this is an evolutionary change, since it leads to a change in the total gene pool of *Homo sapiens*.

Evolution has been characterized as the result of the interplay between mutations (accidental) and natural selection (the determining factor). This is Charles Darwin's theory (so-called Darwinism) in its modern guise—genes and mutations were not known in Darwin's time. Mutations are accidental merely in that they occur without regard for the actual needs of the organism; on the other hand, they cannot occur outside the framework of what is biologically possible. No mutation will lead to the replacement of our skeleton by a steel construction, or change a human being into a dragonfly or even into an ape.

The term *Darwinism* holds that the theory of natural selection is the most important factor in evolution. Formulated in 1859 (actually in 1858) by Darwin, it was immediately embroiled in heated debate. Several alternative theories have been put forth, as, for instance, the inheriting of acquired characteristics: an appealing thought. By acquiring superior characteristics, we might be able to transmit them to our children. But it isn't that easy, since acquired characteristics are not passed on; they must be acquired anew in each generation. Another putative solution was the belief in an inner evolutionary urge, an élan vital, which automatically led the human species forward to ever greater heights. But there is no proof that such an urge exists. Mutations and selection—they are the stuff of evolution, and we must satisfy ourselves with that.

When we look at evolution in broad outline we can disregard mutations and concentrate on selection. Mutations are found in every population, in enormous numbers; all our genetic differences have arisen through mutations. Mutations are the clay; selection is the shaping hand.

Thus we have not—no more than any other organism—developed accidentally, a claim that at times is ascribed to researchers in evolution, with great unjustness. On the contrary: fossil research distinctly reveals marked trends, or evolutionary tendencies, that continue over millions of years. One such trend is the increase in brain size, which can be observed in the history of primates and in that of many other kinds of mammals, such as hoofed animals and beasts of prey. Since increased brain capacity should be advantageous, this development is likely a consequence of natural selection.

Accidents play a part on an entirely different level. We have, for instance, two parallel bones in the forearm, the radius and the ulna. This makes it possible to twist the hand along its longitudinal axis: you can grasp an outstretched hand with the palm up, in which case the two bones are parallel, or turn your palm to face the floor, in which case they lie crosswise. This completely elementary ability is so important—without it all manipulation of tools and other objects would be impossible—that it has to be considered a necessary precondition for our entire development. And this ability is completely dependent on the fact that we have two bones in the forearm—no other type of joint or connection can solve the problem nearly as well.

How did we come to have two bones in the forearm? Fossil finds show that their history goes back to the Devonian period, when fish with tassel-shaped fins lived. (See figure 12.1). Our arms and legs developed from their pectoral and ventral fins. The bone structure in the pectoral fins follows the dichotomous splitting principle: innermost is a solitary bone, then two, then four, etc. (In the outermost smaller bones, which correspond to the bone structure in the wrist and the hand, the exact dichotomy is lost.) Here, the two bones that will become our radius and ulna have no connection with the function they will later serve. The dichotomous branching was perhaps a way to construct a powerful yet elastic fin, but that would hardly have been possible with only one or three

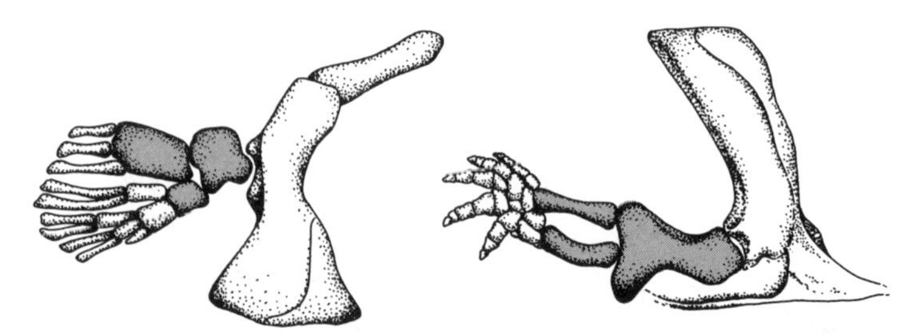

Figure 12.1 The reason we have two bones in the forearm is shown in this sketch of the bone structure of the pectoral fin of a fish with lobed fins from the Devonian period (*left*) and the foreleg of an early amphibian of the Carboniferous period. The homologous bones—the bone in the upper arm, the ulna, and the radius—are shaded. The amphibian's forelegs have developed from the pectoral fin of the fish with lobed fins, an example of "preadjustment."

bones in the prospective forearm. Thus we have an accident, one that at the beginning was a completely unforeseen change in function. There are many similar examples. Our jaws, for instance, are transformed gill arches, and our teeth are transformed dermal scales.

There is a term for this phenomenon, namely *preadaptation,* which implies that an organ may service a com-

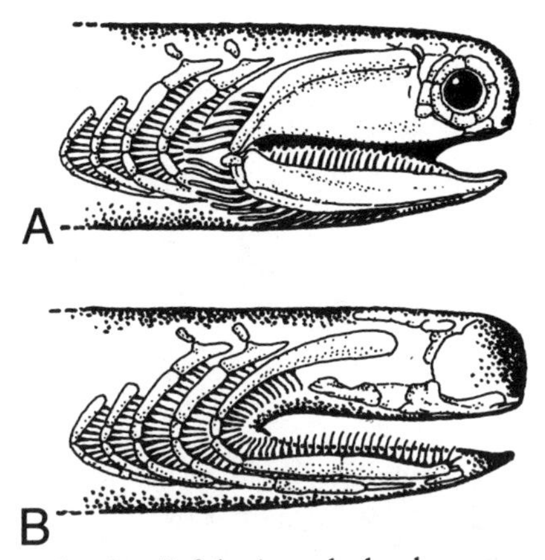

Figure 12.2 The fossil fish *Acanthodes* shows an early stage of jaws developing from a gill arch. (*A*) shows the head of the fish with jaws and gill arches. In (*B*) the jaws have receded and we can see the first actual gill arch, the so-called hyoid arch, which in time gave rise to the bone in our tongue. In humans the gill opening has become partly the ear opening, partly the so-called Eustachian tube leading from the pharynx to the inner ear. Another example of a change in functions.

pletely different purpose than the one for which it originally was developed. Such historical accidents have played a significant role in shaping our destiny, and it is inconceivable that the same course of events would have been repeated otherwise. From this we can conclude that humans are unique in the universe; even though intelligent life may have developed elsewhere, these unknown beings cannot be just like us. If the fish with tassel-shaped fins had had three fins instead of two, perhaps land vertebrates would have been six-legged like the insects. Actually, such a development would have made it easier to assume a half-upright position with the arms free, and such centaurlike beings might have come into existence among several different groups of land vertebrates.

In science fiction, one often reads about so-called humanoids, humanlike beings who have developed on another planet and usually demonstrate an amazing similarity to *Homo sapiens* (with the males usually menacing, and the females seductive). One suspects that centaurlike humanoids are more probable than two-legged ones. Generally speaking, we can presume that intelligent beings developed in different ways. If, for instance, they live in a liquid, such as water, they will probably have a fishlike exterior. Whales, many of which have a larger brain than humans, also seem to distinguish themselves by their high intelligence.

But let us return to the two bones in the forearm. Since they were parts of the first land animals, natural selection could use them and transform them according to the needs of the time. Natural selection functions (as the French biologist François Jacob has maintained) like someone who plods ahead and constructs something or other with the material at hand. The plodder, or tinkerer, is a contrast to the engineer, who creates an exact working drawing, selects the most suitable raw material, and works according to plan with a definitive goal in sight. The engineer attains his result directly. The tinkerer gropes ahead on a winding, wriggling road with constant mishaps and lucky chances, at times with completely unexpected results. That is exactly how the history of fossils appears. We were formed by a tinkerer, and that tinkerer is known as natural selection.

Are present-day human beings subject to natural selection? It has often been said that the progress of medical science has robbed natural selection of its intended function. In Paleolithic times each woman gave birth to many children, but child mortality was exceptionally high and highly selective, which provided the entire population with a genetic disposition for high intelligence and good health. Today, the situation is different. "Bad" genes—those causing nearsightedness, diabetes, difficulty in hearing, etc.—may be compensated for by eyeglasses, insulin, and hearing aids, and thus are not affected by natural selection. On the other hand, it may be said that these genes are no longer bad, since their effects can be prevented. All humanity, as well as our primate relatives, suffers from a genetic shortcoming—we cannot synthesize vitamin C and must receive it through our food or die of scurvy. This is an inheritance from fruit-eating ancestors who had no need of vitamin C synthesis. Is there any difference in principle between obtaining citrus juice, eyeglasses, and insulin? In all these cases a genetic deficiency is compensated for so that a normal life becomes possible.

This theme can be embellished further by pointing out that all human beings in a certain sense have bad genes, that is, genes that do not adapt to all milieus. In a tropical climate, for instance, it is advantageous to have a lanky build and to be dark-complexioned with woolly hair and efficient sweat glands. In a cold climate, on the other hand, it is better to be light-complexioned with a stockier physique and well-developed subcutaneous fat. If these two human types exchange geographical places, their adjustment to their surroundings is reduced—not in any decisive manner but to the extent that natural selection

may begin to function. In more extreme circumstances natural selection will of course function more potently. The physique of Indians in South America's mountains is well adjusted to life in this trying milieu. They have large lungs and hearts, and their blood contains a greater quantity of corpuscles than that of lowland peoples. Although there are many immigrants of other races, both whites and blacks, the Indian type remains unchanged. The immigrants fare badly in the thin air. The number of stillborn children is great, since the fetus does not receive enough oxygen from its mother's blood and consequently suffocates. Natural selection is functioning. And in this case completely normal genes are bad—in *that* particular milieu.

A typical example of a bad gene is the one that causes hemophilia. The life of the hemophiliac is constantly threatened by difficulty in stopping the flow of blood, even from small cuts, and his life is consequently shorter and his progeny fewer than a normal human. It seems, therefore, that there is a strong natural selection process ranged against the hemophilia gene. Still, it doesn't disappear. That is partly because at times a normal gene, through mutation, changes into a hemophiliac gene, and the supply remains filled. But a contributing cause is that the female subject to the disposition (the disease doesn't actually break out in women) has gained increased fertility. A gene that is bad in certain situations may thus be advantageous in other situations. Actually, the phenomenon called *hybrid vigor*—a marked capacity for growth and increased hardiness shown by crossbred plants and animals—implies that if a certain gene brings death when taken in double doses, it brings blessings if it is present in single doses as it is among most people.

That we have robbed natural selection of its function in certain cases does not mean that in other cases it cannot endure and even prosper. One of the most important causes of death in the Western world is traffic accidents. The automobile and the motorcycle have revived the law of the jungle and test our reflexes and our conduct at all times. To the extent that these reflexes depend on genetic factors, natural selection will start to function. The same is true for other forms of violent and sudden death. Even hunger and war may have an effect on natural selection. The soldier in the battle line has usually been selected for his special task—he should be healthy and eager to learn. He is also especially exposed. We have thus created types, subject to natural selection, that may influence human beings of the future—and not always in the way we would wish.

Drugs and other poisons also affect natural selection. We can study their effect on the microbes that produce disease. Certain diseases are cured by antibiotics. In time, the microorganisms develop new, resistant types that are not seriously affected by these substances. New medicines have to be developed and produced, which through selection produce resistant strains, and so on. We tend to develop resistance to poisons in the same way. Alcohol has been consumed for so long that most of us have a moderate resistance to it, but natural selection is still going on, since many people still break down in one way or another; to the degree that such resistance is conditioned genetically (by the condition of the liver, etc.), it reveals ongoing evolution. As far as innumerable new drugs (such as tobacco, marijuana, cocaine, etc.) are concerned, we are merely at the beginning. New drugs are introduced regularly, not to speak of all the other poisons with which we surround ourselves and that are no little threat to nature and thus to us. The situation is too chaotic for us to be able to make any prognoses.

Our foreseeable future is shadowed by many threats—overpopulation, the destruction of the environment, and the atomic bomb. If we can manage to survive them, there is no reason why we cannot look forward to a future spanning billions of years. Our sun is one of the most stable stars in the Milky Way and will provide the earth with just the right dose of heat for a long time. Not to speak of the possibility of our colonizing other heavenly bodies.

It is true that 99 percent or more of all animals have become extinct after having left no progeny. Why should humans escape this (statistically speaking) probable fate? Julian Huxley, on the contrary, has maintained that the human is perhaps the only living being who has the potential for real advances in evolution. All other creatures, according to him, are much too specialized to break out of their set course. A horse cannot become anything but a better horse; something new can evolve only from us. But what that might be remains well hidden.

Our colonizing another heavenly body will doubtless lead to an intensive natural selection process. The same is true, as we have already pointed out, of some of the Earth's most extreme milieus, as, for instance, the mountains of South America. Human settlers on Mars must solve the problems of lack of oxygen and great cold in a technical way but will still be subjected to powerful pressure from natural selection. On Mars the force of gravity is much weaker than on Earth, making it possible to grow into a giant without any greater strain on legs and muscles. And increased size brings many advantages. With greater body size, the proportion between surface and volume is reduced, which implies increased hardiness in frigid temperatures. Furthermore, one needs, relatively speaking, less food and oxygen and can consequently lead a more economical life. When metabolism slows, life span tends to increase. During the few months that a mouse lives, its heart beats as many times as that of an elephant during its seventy years of life. Each of them is living as long as the other—but the mouse is living at a faster pace. Life span is thus subjective, and it would not seem to the

giant human on Mars that he lived longer than the human on Earth.

A colonization of Mars would consequently lead to the birth of a new human race, a race that would not be able to return to Earth without facing enormous problems. Perhaps in the future humans will reach other solar systems. In that case one could expect that each settled planet would in time be inhabited by a unique type of human being. Perhaps in this way Huxley's vision of the future will become reality.

One thing is certain: we and our world are in a state of constant change, in the present as in the past millions of years. There is no brake and there is no turning back. The arrow of time hastens on. We have been amoeba, fish, salamander, reptile, ape, and human being; thus far everything has proceeded without human interference. Now our future lies in our hands.

Glossary

Abbeville. The name of a Stone Age culture most closely associated with *Homo erectus*.

Acheul. The name of a stone-age culture associated with, among others, *Homo erectus*.

Adapidae. Extinct family of prosimians, the predecessors of present-day lemurs, etc.

Agnatha. A fish without jaws; like the present-day lamprey.

Allometry. Unequal rate of growth of different parts of the body. Example: the brain grows more slowly than the rest of the body.

Amphibian. A batrachian, an animal that can live in the water and on land.

Artifact. A man-made object, in contrast to any found in nature; if from the Ice Age, usually a stone object used as a tool, etc.

Brachiation. Swinging by the arms from branches of trees.

Catarrhini. Narrow-noses; includes humans, apes and the Old World monkeys.

Darwinism. The theory of natural selection.

Dichotomonous splitting. Continuous splitting into two parts.

Eocene. The second epoch of the Tertiary period; thirty-five to fifty-five million years before our time.

Eon. Geologic time period comprising several eras.

Epoch. Geologic time sector; part of a period.

Era. Long time sector comprising several periods.

Eucaryote. Cell with a nucleus.

Gene pool. The sum of hereditary aptitudes within a population.

Holocene. The geological period encompassing the present time; the last ten thousand years.

Hominidae. Family comprising our genus *Homo* human and its closest relative *Australopitheas*.

Hominoidea. Superfamily including the families Hominidae, Pongidae, Hylobatidae, and the extinct Pliopithecidae.

"Humanoid." Science fiction term for humanlike beings in space.

Hylobatidae. Family comprising the gibbons and their extinct relatives.

-idae. Ending of names for zoological families.

Miocene. The fourth epoch of the Tertiary period; twenty-five to five million years before our time.

Monophyletic group. A group that represents only one lineage, including all its descendants.

Mousterian culture. The typical Stone Age culture of Neanderthal man.

Mutation. A change in the hereditary composition of a being.

Neoteny. Embryonic or "childish" characteristics remaining in adults.

-oidea. Ending of names for zoological superfamilies.

Oldowan. The name of a primitive Stone Age culture associated with *Homo habilis*.

Oligocene. The third epoch of the Tertiary period; thirty-five to twenty-five million years before our time.

Omomyidae. Extinct prosimian family, predecessors of the present-day tarsier, probably also of the higher primates.

Paedomorphosis. See *neoteny*.

Paleocene. The first epoch of the Tertiary period, sixty-five to fifty-five million years before our time.

Paleolithic period. The Old Stone Age; began 2 to 2.5 million years ago and ended about the end of the Ice Age.

Paleontology. The study of ancient organisms.

Paraphyletic group. A group that contains parts of two or more lines of development.

Parapithecidae. Extinct family of primates living in the Oligocene period.

Period. Geological time sector between era and epoch.

Pharynx. The section of the throat located between the vocal cords and the oral cavity.

Phylogeny. Evolution; the development of type or family.

Platyrrhini. Broadnoses; comprises the New World monkeys.

Pleistocene. The first epoch of the Quaternary period (the Ice Age); 1,600,000 to 10,000 years before our time.

Plesiadapiformes. An extinct group of prosimians.

Pliocene. The fifth epoch of the Tertiary period; 5 to 1.6 million years before our time.

Pongidae. Family comprising the anthropoid apes: gorillas, chimpanzees, and orangutans, in addition to their extinct relations.

Primates. An order of mammals comprising prosimians, monkeys, apes, and human beings.

Procaryotes. Cell without a cellular nucleus (for instance, bacteria).

Protein. Albuminous substance in plants and animals.

Radiometry. Measuring geologic time by means of the rate of disintegration of radioactive elements.

Tarsus, tarsal. Relating to small bones of the foot of a vertebrate, between the metatarsus and the tibia.

Taxonomy. Orderly classification of plants and animals.

Bibliography

Fichtelius, K. E. 1985. *Hur apan miste pälsen och kom upp på två ben* (How the Ape Lost Its Fur and Got Up on Two Legs). Akademilitteratur.

Heintz, A. and N. Heintz. 1968. *Människans ursprung* (The Origin of Man). W. & W.

Howells, W. W. 1964. *Mankind in the Making: The Story of Human Evolution*. Heinemann.

Jelinek, J. 1978. *Den stora boken on människans forntid* (The Big Book About Man's Past). Tiden.

Johanson, D. C., and M. A. Edey. 1981. *Lucy: The Beginning of Humankind*. Simon and Schuster.

Koenigswald, G. H. R. von. 1957. *Möten med urtidsmänniskan* (Meeting with Ancient Man). Natur och Kultur.

Koenigswald, G. H. R. von, ed. 1958. *Hundert Jahre Neanderthaler* (A Hundred Years of the Neanderthals). Böhlau Verlag.

Kurtén, B. 1988. *Singletusk: A Novel of the Ice Age*. J. Curley.

Kurtén, B. 1972. *The Age of the Mammals*. Columbia University Press.

Leakey, R. E. and R. Lewin. 1982. *Origins: What New Discoveries Reveal About the Emergence of Our Species and Its Possible Future*. Dutton.

Lewin, R. L. 1987. *Bones of Contention*. Simon and Schuster.

Maglio, V. J., and H. B. S. Cooke, eds. 1978. *Evolution of African Mammals*. Harvard University Press.

Morris, D. 1970. *The Naked Ape*. Dell.

National Geographic. November 1985.

Natural History. Ancestors series, 1984.

Reader, J. 1981. *Missing Links: The Hunt for Earliest Man*. Viking Penguin.

Scientific American. June 1983 and March 1984.

Stephansson, O. 1983. *Sveriges vandring på jorden* (The Wandering of Sweden on the Earth). Liber.

Szalay, F. S. and E. Delson. 1979. *Evolutionary History of the Primates*. Academic Press.

Wood, B. 1976. *The Evolution of Early Man*. Eurobook.

Index

Italicized page references signify illustrations.

Designer: Teresa Bonner
Text: Galliard
Compositor: Maple Vail
Printer: Maple Vail
Binder: Maple Vail